Computational Techniques in Physics

QMW LIBRARY
(MILE END)

Computational Techniques in Physics

P K MacKeown and D J Newman
Department of Physics,
University of Hong Kong

Adam Hilger, Bristol

British Library Cataloguing in Publication Data

MacKeown, P.K.
Computational techniques in physics.
1. Physics—Data processing
I. Title II. Newman, D.J.
530.1′594 QC52

ISBN 0-85274-537-0
ISBN 0-85274-548-6 Pbk

A software package to accompany Computational Techniques in Physics is available from Adam Hilger. This is for use with selected projects and contains programs in BASIC and TURBO PASCAL, as well as a graphics program and data files. The software is suitable for running on IBM-PC microcomputers and compatibles. For more information and an order form, please see the back of the book.

Published under the Adam Hilger imprint by IOP Publishing Ltd
Techno House, Redcliffe Way, Bristol BS1 6NX, England

Printed in Great Britain by J W Arrowsmith Ltd, Bristol

Contents

*Sections marked with an asterisk could be omitted in an introductory course.

Preface

This book is largely the outcome of teaching an elective course in computational physics over the past several years to classes of final year undergraduate students at the University of Hong Kong, although some of the material originated more than 20 years ago in a course taught at Queen Mary College, London. Even allowing for limitations dictated by such a constituency, much of the material may also be of interest to physics postgraduate students who have not been exposed to such a course, as well as to some students of physical chemistry and materials science. Only exposure to basic physics courses, and the mathematical level commonly required for such courses, together with a familiarity with elementary computer programming such as is common today, is assumed. For convenience, the main mathematical requirements, namely the properties of matrices and basic concepts in statistics, are summarised in the Appendix at the end of the book.

The variety of topics available to such a course is very great. In our choice we have tried to emphasise, within the time available, techniques of wide applicability in physics which are not so readily available in texts at the undergraduate level. This has meant the exclusion of several not unimportant topics, such as quadrature, ordinary differential equations and optimisation, which are widely described, as well as commonly provided for in library packages. It has also meant the omission, with perhaps less justification, of several important current topics in computational physics such as symbolic computation, fractals, chaos etc.

Even with these omissions, there is more material available than could be mastered in the amount of time usually allotted to such a course. To allow for this, beyond Chapter 1 some flexibility exists in the presentation so that a selection can be made from among the topics.

Logical connections between chapters are summarised in the following diagram:

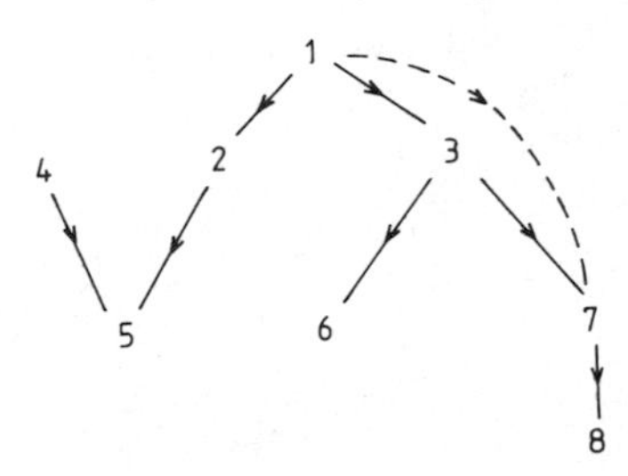

This shows three basic strands, leading to fitting of parametrised matrices to eigenvalues (Chapter 5), the finite element method (Chapter 6) and the Monte Carlo method (Chapters 7 and 8). Note that Chapter 3 is not essential to the understanding of Chapters 7 and 8. Sections containing more advanced material are marked with an asterisk (*) in the contents list and could be omitted in an introductory course.

From the technical point of view, much of our students' work was performed in a departmental microcomputer laboratory (sometimes augmented by using home personal computers). Generally, preprepared material was made available in BASIC and PASCAL but no rigid stipulation as regards programming language was made; FORTRAN programs also turned up in reports submitted by the students on some projects. In the same spirit, we give no explicit program listings in this book; details of some published listings are mentioned at relevant places in the text. In addition, a software package (*Software for Computational Techniques in Physics* by P K MacKeown, D J Newman and M F Reid) including a diskette containing programs for selected projects in BASIC and TURBO PASCAL, as well as a graphics program and data files, suitable for running on IBM-PC microcomputers and compatibles, is available from Adam Hilger (see the back of the book for more details and an order form).

Acknowledgments

It is difficult to select any one person from among the very large number of people who have helped the authors to develop the projects presented in this book for special acknowledgment. However, we should particularly like to mention Miss Catherine Lau, who assisted the first batches of students who took our course in Hong Kong, and Miss Flora Yip who wrote some excellent programs for our microcomputers. We are also grateful to Dr Mike Reid for his comment on draft chapters of this work.

We gratefully acknowledge permission to reproduce material as follows: for figure 1, the International Atomic Energy Agency in Vienna; for figure 2.2, Appendix 2A and table 4.1, *The Journal of Chemical Physics*; for Appendix 2B, *The Journal of the Physics and Chemistry of Solids*; for Appendix 4A, the *American Journal of Physics*. The copyright for material from the *Journal of Chemical Physics* and the *American Journal of Physics* is owned by the American Association of Physics Teachers.

P K MacKeown
D J Newman
May 1986

Introduction

The Scope of Computational Physics

Computational methods have become established as a research tool in physics in a way that is barely hinted at in standard textbooks on, say, quantum mechanics or statistical mechanics. Most elementary texts on quantum mechanics, for example, give pride of place to analytical solutions for the hydrogen atom and harmonic oscillator with scarcely a mention of the power of the Hartree–Fock self-consistent approach in the generation of realistic one-electron solutions for many-electron atoms.

One of the difficulties encountered with analytical methods in theoretical physics is that they are very problem specific, so that having solved one problem the challenge of a problem in a related field is still very great. In addition, they are often inflexible in the sense that even a small variation in the original problem may mean the methods previously used are inapplicable. For example, although the (one-dimensional) Schrödinger radial wave equation can be solved analytically for specific potentials, the work involved in solving it for a slightly different potential will be at least of the same order, and often very much greater. A second difficulty may arise from the complexity of the problem. Although a formal solution may be written down, numerical outputs may not be obtainable without introducing approximations in the evaluation and it is at this final stage that the physical significance of such approximations is least apparent. A graphic illustration of this limitation arises in the case of the pair of coupled integro-differential equations which describe the extent of the cascade of electrons and photons which builds up when a high energy electron is incident on a block of material. The varying role of the many competing physical processes, e.g. Coulomb scattering, Compton scattering, bremsstrahlung, pair production and ionisation, throughout the cascade results in a theoretical formalism which is too complex to evaluate with an accuracy sufficient for comparison with experimental results. With the advent of high speed computers it is better to simulate the whole process, i.e. imitate the behaviour of each particle in the cascade individually. Such simulation enables not only the average behaviour over many cascades to be determined but also the intercascade fluctuations, an aspect which is almost impossible to solve by analytical methods.

Partly as a result of the current concentration on 'neat' analytic solutions in their training, many theoretical physicists have come to regard computational methods as a necessary evil in research, to which they may be forced to resort when analytical ingenuity fails them. We believe, however, that the computational approach has an essential unity of its own, making a given method applicable to a far wider range of problems than any analytical approach. For example, in the study of a complex system of competing processes, such as that described above, the computational approach is to start with the primitive elements of the system (i.e. the interactions) and to 'build up' a solution. Such a 'modelling' procedure can be readily adapted to other problems of a similar type. A training in computational physics will thus provide the student with tools of considerable flexibility in attacking problems in different areas of physics.

It must not be thought, however, that any problem can be solved simply by inserting a representation of its most primitive elements into a computer and cranking a handle. Such an approach would probably take an inordinate amount of time and the output would probably not be capable of interpretation. Methods of casting of the problem into a suitable form, e.g. a set of differential equations, to serve as a basis for the computation form a major aspect of the computational approach. Illustrations of this may be seen in the four-colour problem in pure mathematics and the Kondo problem in solid state physics. The former, which states that four colours are sufficient for colouring any plane map, had been formulated for 200 years before its solution by computer in 1976. The Kondo effect, which is a characteristic behaviour of the temperature dependence of the resistivity of a metal containing magnetic impurities, was extensively studied analytically before a formalism was obtained which lent itself to solution by computational methods in 1975 (since which time an analytical solution has also been obtained). Apart from explaining observed phenomena by elucidating the physical content of a complicated theoretical formalism, these methods may also predict new phenomena not obvious from the analytical formalism. For example, the investigation of behaviour as a function of temperature when different intermolecular potentials are modelled in a system with a large number of molecules may reveal a possible phase transition in the system.

At another level, modelling procedures may be regarded as a substitute for experiment. They provide a method of approximating experiments which cannot readily be carried out in the laboratory, either because of their dimensions or the time required or because of the cost involved. We may cite examples in astrophysics and semiconductor electronics. Astronomers need no longer be purely passive observers of the universe. Since most of the basic physical laws are believed known, they may simulate controlled experiments on a cosmic scale, e.g. the evolution of galaxies, following their development over times of the order of the age of the universe. In the study of the electronic properties of different combinations of semiconductors there are many possible systems and the conditions under which laboratory experiments must be carried out are exacting. It may therefore be more economical to simulate their properties in order to decide in advance which systems are most promising before embarking on laboratory experiments. Many other examples in the fields of plasma physics, atomic physics, statistical physics etc can be found in the literature.

Yet another aspect of computer modelling is the direct testing of physical hypotheses against experimental data. An important example of this application is the development of parametrisation schemes to fit spectroscopically determined energy spectra. Any such scheme, in order to be physically significant, must be built on specific hypotheses concerning the nature of the system under study. In so far as the number of data exceeds the number of parameters, it is possible to test these hypotheses by the least squares fitting procedures, both the quality of the fit and the values of the fitted parameters providing useful criteria regarding the accuracy of the hypotheses. If a given hypothesis is found to be consistent with the experimental data it may be safely built into first principles calculations.

A simple example of a parametrisation scheme is provided by the potential energy expression for the electrons in the partially filled shells of paramagnetic ions in crystals. It is possible to test the hypothesis that the anisotropic, or crystal field, part of this energy can be broken down into a sum of one-electron terms. Such a crystal field model is found to work very well, so that first principles work can be directed to calculating contributions to the crystal field parameters, rather than calculating the observed energy levels directly.

The relationship between computational physics and the experimental and theoretical branches of the subject, of which we have given some examples above, is more generally and concisely summarised in figure 1.

The aim of this book is to describe the computational approach to problems in theoretical physics. This requires the student to have an adequate mathematical background, indicated by the material summarised in the Appendix. It also requires an interest in many aspects of physics, for students will get the most out of this book if they delve more deeply into the physics underlying some of the projects.

In carrying out the projects, use can be made, where appropriate, of the standard numerical analysis packages that are available for most computers. Sufficient information is given, however, to enable students with some programming expertise to construct their own programs for the projects.

All the projects given can be carried out using standard 'personal' microcomputers with disc drives and printer. They have mostly been tested by several groups of students over the last five years and the experience gained from this is reflected in the project descriptions.

A useful review of scientifically orientated program packages has recently been given by J C Nash (1985). A comprehensive compilation of computational algorithms for standard numerical methods (specially designed for use on microcomputers) which is recommended is Press *et al* (1986). Students looking for further projects will find many suggestions in the American

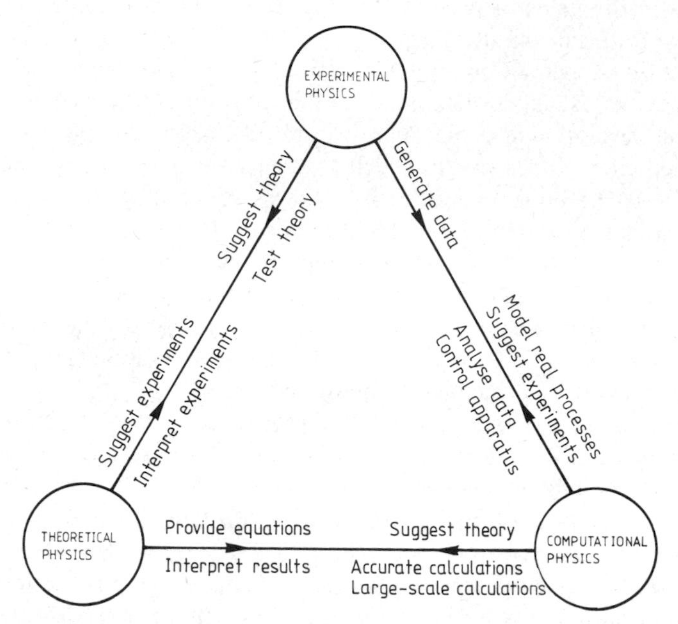

Figure 1. The relationship between the three aspects of physics, indicating the coupling of computational physics to both experimental and theoretical disciplines (after Roberts (1973)).

Journal of Physics (for example, Bahurmuz and Loly (1981) for electron band structure calculations and van der Merwe (1980) for a discussion of electron optics). Both *Computer Physics Communications* and the Quantum Chemistry Program Exchange (Department of Chemistry, Indiana University, Bloomington, Indiana, 47405, USA) supply special purpose program packages at very reasonable prices, and the more ambitious student is sure to find something of interest in their catalogues.

Accuracy of Computational Methods

What we have said up to now has emphasised that the content of the subject is very much more physics than computing, but at the same time we cannot avoid the fact that the technical procedures rely heavily on *numerical methods*. Since these techniques are only touched upon in conventional mathematics courses for physicists it will be necessary to devote some space to them. In practice, however, most of the standard procedures, such as finding eigenvalues or integrating, are available in standard software packages. It is therefore often unnecessary to know more than the general principles involved. Hence, we shall avoid excessive detail. There is one aspect however, namely the accuracy with which a numerical solution may be obtained, which will be briefly examined here.

It will be fairly obvious that numerical methods cannot give an 'exact' solution to a problem, but then in physics we never require 'exact' solutions, only solutions which are more precise than the appropriate experimental data. It follows that we must have some idea of the uncertainty in the experimental data and we must be aware of the uncertainty in any result obtained by computation. Two sources of uncertainty occur in computed results, arising respectively from (i) inherent approximations in the different numerical methods used, often called *truncation errors*, and (ii) approximations in the machine evaluation of the result due to the limited accuracy with which numbers are stored and manipulated by the computer, known as *round-off errors*. Both sources of error can be minimised by using careful numerical techniques.

We consider first the way in which round-off errors occur. Non-integers are held in a normalised floating point representation, generally to a base β in the form

$$0 \cdot d_1 d_2 \ldots d_k \times \beta^m \qquad (d_1 \neq 0,\, d_i < \beta).$$

For example, using base 8

$$(9.15)_{10} = (11.11463146314\ldots)_8 = 0.1111463146314\ldots \times 8^2.$$

The word size is k and digits above k are not retained, i.e. the figure is rounded off to k digits; how this is done varies from machine to machine. If, in the above example, $k = 6$, too small a value in practice, the number processed is $0.111146 \times 8^2 \simeq (9.1499023438)_{10}$. We see that the totality of numbers in a finite interval accessible to the machine is finite. During execution the element amongst them which is just less than the number we wish processed will probably be used. Because of this non-identity of the numbers manipulated with the elements in the theory, the results will not be exactly equal to those from an ideal computer; as an example we may cite the failure of commutativity, $a + b + c \neq c + b + a$, in general.

With modern computers, the number of equivalent decimal digits in the mantissa of each number will be quite large, of the order of 10–16, so the effect of round-off in any individual operation is not likely to be a serious cause for concern. However, the cumulative effect in an extended calculation may lead to a significant error, in the case of a simulation possibly giving rise to an irreversibility which is not characteristic of the input model. Some elementary precautions should therefore be taken to avoid such effects. Some computers provide the facility of *double precision* which enables specified quantities to be represented and manipulated using words with twice as many bits as normal, this approximately doubles the number of significant decimal digits. It is difficult to give general prescriptions for minimising the relative round-off error. However, one can say, in respect to addition, that it will be least when quantities of similar magnitude are being added, hence the importance of the nested arrangement (Horner's rule) for evaluating a polynomial:

$$\sum_{0}^{n} a_j x^j \to a_0 + x(a_1 + x(a_2 + \cdots + x(a_{n-2} + x(a_{n-1} + xa_n))\ldots)).$$

On the other hand, in subtraction, just this situation should be avoided, since if the only difference between two numbers is in the last digits of their rounded representation the relative error may be large. The construction of Bessel functions using recurrence relations has been found to be particularly susceptible to this type of error.

The results of using any numerical method are almost invariably approximations to the function being evaluated. The truncation error is the discrepancy between the true (usually unknown) value of the function $f(x)$ and the exact estimate $\tilde{f}(x)$ given by the numerical methods, that is

$$\varepsilon(x) = |f(x) - \tilde{f}(x)|.$$

This error is additional to any error arising from round-off. In this case also, the error contributed in any single calculation may be made insignificant; however, care needs to be taken in iterative type operations to avoid accumulation of a significant error and ensure convergence to a solution. An elementary example of a truncation error occurs in approximating the derivative of the function $f(x)$ by

$$f'(x) = (f(x+h) - f(x))/h.$$

The truncation error here

$$\varepsilon(x) = \frac{h}{2!} f''(x) + \frac{h^2}{3!} f'''(x) + \cdots$$

is said to be of order h, $\varepsilon \sim O(h)$. It will be considered further in Chapter 3.

References

Bahurmuz A A and Loly P D 1981 *Am. J. Phys.* **49** 675
van der Merwe J P 1980 *Am. J. Phys.* **48** 569
Nash J C 1985 *Byte* **10** 145–50
Press W H, Flannery B P, Teukolsky S A and Vetterling W T 1986 *Numerical Recipes—The Art of Scientific Computing* (Cambridge: Cambridge University Press)
Roberts K V 1973 in *Computing as a Language of Physics* (Vienna: IAEA)

Chapter 1

Linear Fitting and Interpolation Using Transformation and Least Squares Methods

In this chapter we study the mathematical aspects of fitting parameters to data using direct transformation and linear least squares methods. Transformation methods are most appropriate in cases where the data have negligible scatter, so that a curve can reasonably be fitted through all the given points. Minimisation of the sum of squared deviations is appropriate when the data are too scattered to allow a simple curve fitting. It is not, of course, the only possible method of determining the 'best fit' parameters but does, nevertheless, have some rather nice mathematical properties which have put it into a pre-eminent position.

1.1 Interpolation by Transformation

Given N data of the form (x_i, y_i) with negligible errors we may fit a linear expression

$$y_i = \sum_{\alpha=1}^{N} a_\alpha f_\alpha(x_i) \tag{1.1}$$

through all the data points, provided that the set of N vectors f_α $(\alpha = 1, \ldots, N)$ are independent, so that a solution to the N simultaneous equations for the parameters a_α may be found. This procedure enables us to estimate the values of y between data points (interpolation) and outside the range of data points (extrapolation). A common choice for the $f_\alpha(x_i)$ are the sine and cosine functions, in which case the expansion (1.1) is called a Fourier series. These functions have the very useful property that they provide both an infinite set of orthogonal functions over a continuous (finite) range of x and a finite set of orthogonal functions over any finite and evenly distributed set of x points, $x_i = \mu i + \nu$ $(i = 1, \ldots, N$; μ and ν constant). With this specific example in mind, let us suppose that the $f(x_i)$ are chosen to be orthogonal over the range of the data points x_i:

$$f_\alpha^T f_\beta = \sum_i f_\alpha(x_i) f_\beta(x_i) = 0 \qquad \text{if } \alpha \neq \beta. \tag{1.2}$$

It is then possible to invert equations (1.1) simply by multiplying both sides by $f_\beta(x_i)$ and summing over i. We obtain

$$a_\beta = \sum_i y_i f_\beta(x_i)\left(\sum_i f_\beta^2(x_i)\right)^{-1} = \boldsymbol{y}^T \boldsymbol{f}_\beta / \boldsymbol{f}_\beta^T \boldsymbol{f}_\beta. \tag{1.3}$$

Hence, in this ideal case, there is a simple solution of the simultaneous equations giving the parameters a_β directly in terms of the data points y_i.

In some circumstances it may be advantageous to use a transformation method, even when data errors cannot be neglected. This is the case, for example, in the conversion of interferometric data (amplitude against time) into spectra (amplitude against frequency or wavelength) by Fourier transformation in analysing the output of interferometers (Strong and Vanasse 1959). In this way, we ensure that no information is lost, as it inevitably is when least squares fitting procedures are used. Fourier transformation methods are discussed in §1.4.

1.2 Polynomial Interpolation

It is always possible to fit a unique polynomial (equation (1.1) with $f_\alpha(x) = x^{\alpha-1}$) of order $(N-1)$ to N data points (x_i, y_i) using the transformation method. This may be written as

$$P^{(N-1)}(x) = \sum_{i=1}^{N} y_i N_i(x)$$

where

$$N_i(x) = \prod_{j=1, j\neq i}^{N} \frac{(x - x_j)}{(x_i - x_j)}. \tag{1.4}$$

For the case where the function $y(x)$ generating the data points is itself a polynomial of order $v < N$, $P^{(N-1)}(x)$ will coincide with y everywhere. However, in general, $P(x)$ coincides with $y(x)$ only at the data points and not at intermediate points, nor does the derivative $P'(x)$ coincide with $y'(x)$, even at the data points. An example of a particularly unsatisfactory fit by a polynomial is shown by the broken curve in figure 1.1, where a tenth-order polynomial has been fitted to 11 equally spaced values of the physically important function $y(x) = 1/(1 + x^2)$†. Even for cases where such oscillatory behaviour would not occur, it is often not possible to fit a higher order

†This function, known as the Lorentz or Breit–Wigner distribution, occurs frequently in physics, representing as it does, in the absence of noise, the intensity profile of a spectral line or the cross section for elastic scattering of a particle in the neighbourhood of a resonant energy.

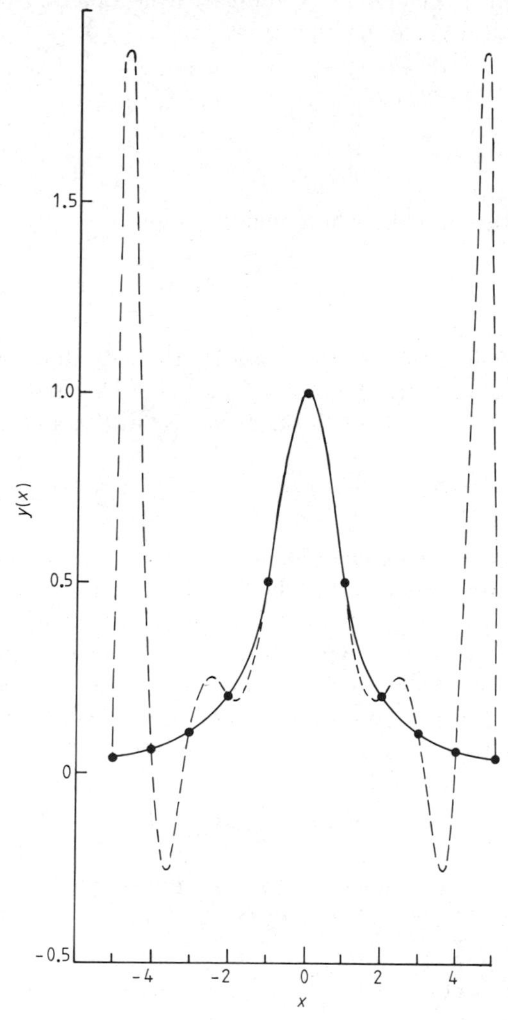

Figure 1.1 Interpolation based on 11 given values of the function $y(x) = (1 + x^2)^{-1}$ denoted by the solid dots. The broken curve is a tenth-order polynomial fitted to the points using (1.4). The solid curve is a cubic spline fitted to the same points, see §1.3.

polynomial because of computational difficulties, very small determinant values leading to overflow etc. Many methods, often including transformation of the independent or dependent variable, are discussed in books on numerical analysis.

Many of the defects of 'global' polynomial interpolation, as discussed above, can be removed by using 'piecewise' polynomial interpolation. Given a set of N nodal points (x_i, y_i) defining a set of $N - 1$ intervals $(x_i - x_{i+1})$, subsequently called 'elements', we represent the function by a polynomial $P_i(x)$ in the interval (x_i, x_{i+1}), that is

$$P_i(x) = \sum_{j=0}^{m} a_{ij} x^j \qquad x_i < x < x_{i+1} \tag{1.5}$$

$$= 0 \qquad x \notin (x_i, x_{i+1}).$$

We can then write over the whole region

$$y(x) = \sum_{i=1}^{N-1} P_i(x). \tag{1.6}$$

These polynomials must at least satisfy the continuity conditions, i.e. $P_1(x_1) = y_1$, $P_{i-1}(x_i) = P_i(x_i) = y_i$, $i = 2 \ldots (N-1)$, and $P_{N-1}(x_N) = y_N$. The simplest case is $m = 1$, or linear piecewise fitting, in which

$$P_i(x) = \frac{(x_{i+1} - x)y_i + (x - x_i)y_{i+1}}{x_{i+1} - x_i} \tag{1.7}$$

where $x_i \leqslant x \leqslant x_{i+1}$. This method, which simply joins adjacent points by straight lines, is very crude in that the first derivative of the fitted function is discontinuous at each nodal point and would seldom be used. A more practical version is to use polynomials of order three on each interval, which ensures continuity of the first and second derivatives at the nodal points. Such an interpolation function, known as a *cubic spline*, will be discussed shortly.

Inserting equation (1.7) into (1.6) we can write

$$y(x) = \sum_{i=1}^{N-1} y_i N_i(x)$$

which defines the basis functions for the interpolation or *shape functions* $N_i(x)$, as shown in figure 1.2. For $i > 1$

$$N_i(x) = \frac{x - x_{i-1}}{x_i - x_{i-1}} \qquad x_{i-1} \leqslant x \leqslant x_i$$

$$= \frac{x_{i+1} - x}{x_{i+1} - x_i} \qquad x_i \leqslant x \leqslant x_{i+1}.$$

Linear interpolation is of somewhat greater value when we are dealing with a function of two or more variables, e.g. $z(x, y)$. For example, if values of the function are known at three distinct points, $z_i = z(x_i, y_i)$, $i = \alpha, \beta, \gamma$, we can approximate $z(x, y)$ at any point (x, y) in the neighbourhood of the points in terms of z_α, z_β and z_γ in a manner analogous to (1.7).

Let us write $P(x, y) = a_0 + a_1 x + a_2 y$. The requirement that $P(x_i, y_i) = z_i$, $i = \alpha, \beta, \gamma$, gives us three equations which determine the a_i values, i.e.

$$\mathbf{R}\,\mathbf{a} = \mathbf{z}$$

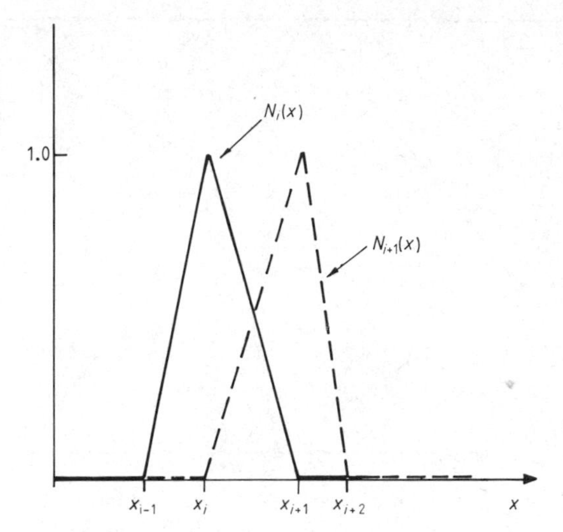

Figure 1.2 Examples of the shape functions occurring in piecewise linear interpolation in one dimension.

where $\mathbf{a}^{\mathrm{T}} = [a_0 a_1 a_2]$, $\mathbf{z}^{\mathrm{T}} = [z_\alpha z_\beta z_\gamma]$ and

$$\mathbf{R} = \begin{bmatrix} 1 & x_\alpha & y_\alpha \\ 1 & x_\beta & y_\beta \\ 1 & x_\gamma & y_\gamma \end{bmatrix}$$

The solution is,

$$\mathbf{a} = \mathbf{R}^{-1}\mathbf{z}$$

and hence can be written, in analogy with (1.7), as

$$P(x, y) = N_\alpha(x, y)z_\alpha + N_\beta(x, y)z_\beta + N_\gamma(x, y)z_\gamma. \tag{1.8}$$

Here the linear functions in x and y, given by

$$N_\alpha(x, y) = \frac{1}{2A}[(x_\beta y_\gamma - x_\gamma y_\beta) + (y_\beta - y_\gamma)x - (x_\beta - x_\gamma)y] \qquad \alpha, \beta, \gamma \text{ cyclic} \tag{1.9}$$

are the shape functions for the triangle. A is the area of the triangle defined by the three points and arises from the relation det $\mathbf{R} \equiv 2A$. In this way a function can be interpolated over a plane in terms of its values specified at the vertices of triangles (elements) where functions $P_\mu(x, y)$ of the form (1.8) are defined in each triangle labelled $\mu = 1, 2 \ldots$, as in figure 1.3, and zero elsewhere. That is,

$$z(x, y) = \sum_\mu P_\mu(x, y).$$

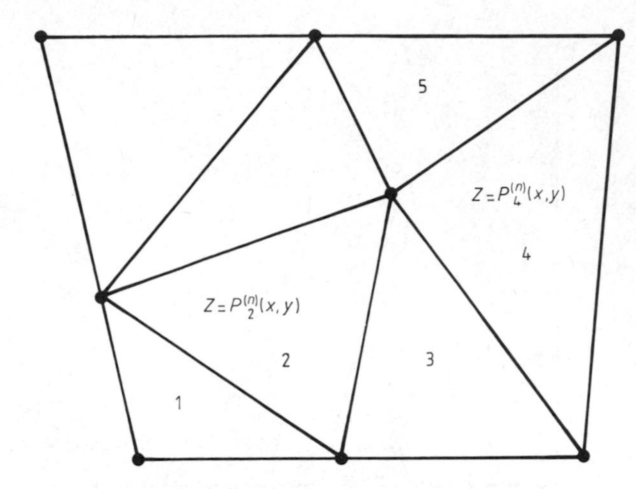

Figure 1.3 An illustration of piecewise polynomial interpolation in two dimensions of a function $z(x, y)$. In a segment numbered μ the function is approximated by an nth-order polynomial, $P_\mu^{(n)}(x, y)$, whose coefficients are determined by the given values of the function at the nodes.

At the common edge of two triangles this function is continuous (because there is a unique linear function passing through the two vertices common to each of the triangles). However, its first derivatives are discontinuous.

This piecewise interpolation in two (or more) variables is the starting point for the important method of finite elements in the numerical solution of partial differential equations which we shall discuss in Chapter 6.

1.3 The Cubic Spline

Returning to the case of a single independent variable, we wish to write $y(x)$ in terms of $(N-1)$ polynomials, as in equation (1.6). If we choose cubics, i.e. $m = 3$ in equation (1.5), we are able to ensure that the function y is not only continuous at each of the nodes but also has continuous first and second derivatives. Setting $P_i(x) = \Sigma_{j=0}^{3} a_{ij} x^j$ means that we have to solve for $4(N-1)$ coefficients a_{ij}. The conditions for continuity give us $2(N-1)$ equations:

$$P_i(x_i) = y_i \qquad P_i(x_{i+1}) = y_{i+1} \qquad i = 1, \ldots, (N-1).$$

The requirement of continuity of the first and second derivatives at the interior nodes gives an additional $2(N-2)$ equations:

$$P'_{i-1}(x_i) = P'_i(x_i) \qquad P''_{i-1}(x_i) = P''_i(x_i) \qquad i = 2, \ldots, (N-1).$$

This total of $4N - 6$ equations is still two short of the number of unknowns and two additional conditions must be specified in order to solve for the a_{ij}. In the usual case, where further information is lacking, it is usual to approximate the second derivative at the extreme points to zero, i.e. to assume that the curve continues outside each end of the region with constant slope:

$$P_1''(x_1) = 0 \qquad P_{N-1}''(x_N) = 0. \tag{1.10}$$

The resulting curve, which can be shown to be the curve of least curvature passing through the N points, is known as a natural cubic spline.

The cubic spline can be constructed in the following way.

(i) Define $h_i = x_{i+1} - x_i$, $d_i = (y_{i+1} - y_i)/h_i$ for $i = 1, \ldots, (N-1)$.

(ii) Solve the following N simultaneous equations for the coefficients c_i, $i = 1, \ldots, N$:

$$c_1 + 0 \cdot c_2 = 0$$

$$h_1 c_1 + 2(h_1 + h_2)c_2 + h_2 c_3 = 6(d_2 - d_1)$$

$$h_{i-1}c_{i-1} + 2(h_{i-1} + h_i)c_i + h_i c_{i+1} = 6(d_i - d_{i-1})$$

for $i = 3, \ldots, (N-1)$ and

$$0 \cdot c_{N-1} + c_N = 0.$$

To evaluate the function at the point x we must find the interval in which x falls, e.g. $x_j < x < x_{j+1}$, and define the local variable $t \equiv (x - x_j)/h_j$, then we can write

$$y(x) = P_j(t) = y_j + \frac{h_j^2}{6}\left[(c_{j+1} - c_j)t^3 + 3c_j t^2 + \left(\frac{6d_j}{h_j} - 2c_j - c_{j+1}\right)t\right]. \tag{1.11}$$

From equation (1.11) the first two derivatives of the interpolated function can be readily obtained. While it is not easy to say, because of ignorance of the validity of approximation (1.10), just how good a spline fit will be, it is often very good. In figure 1.1 it is compared with the tenth-order polynomial in the fit to $1/(1+x^2)$ and does not diverge from the function by more than 3%. In figure 1.4, we show a spline fit to a complicated function in the theory of electromagnetic cascades, $\lambda_1(s)$, and compare its derivative with known values of the derivative $\lambda_1'(s)$. Despite the poor approximation (1.10), at $s = 0.1$, $\lambda_1''/\lambda_1 \simeq 100$; both the function and its first derivative are well represented away from the end points.

More sophisticated versions of spline functions are discussed in texts on numerical analysis. Explicit FORTRAN programs for calculating the simple cubic spline are given in Cheney and Kincaid (1980) and in Stroud (1974).

Before closing this section on interpolation a warning is in order. In practice, interpolation will often be an intermediate step in the execution of a larger program, where the accuracy of the fit will not be inspected at every stage. Although a certain truncation error will be tolerable in such a calculation, it is important that physical requirements are not violated, because even small violations may give rise to wholly spurious results. As an example we can cite the tenth-order polynomial fit to the Lorentz distribution in figure

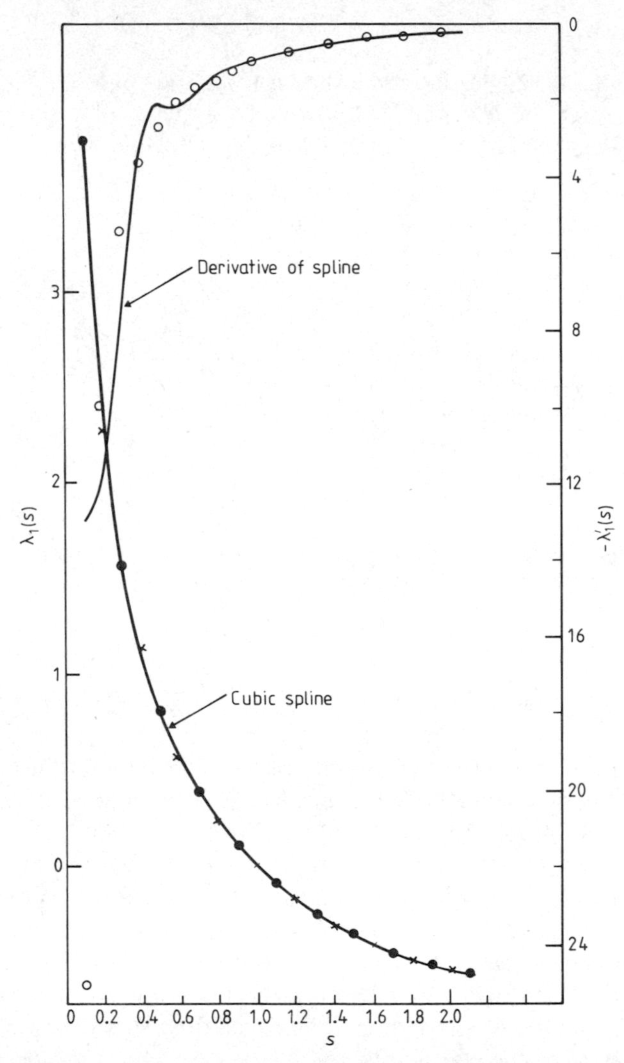

Figure 1.4 Interpolation using a cubic spline on the basis of 11 values of the function $\lambda_1(s)$, shown by the solid dots. The accuracy of the fit may be judged by comparing it with additional known values of the function, shown by crosses. The derivative of the interpolating spline (upper curve) is also compared with known values of $\lambda_1'(s)$, shown by open circles. (The origin of the function $\lambda_1(s)$ may be found in Rossi and Greisen 1941.)

1.1; if the data represented spectral intensities, the negative excursions in the interpolated curve would have no physical interpretation. A more serious case occurs when interpolating a magnetic field, as is often necessary in the simulation of plasmas. Even though the interpolated field $\tilde{\boldsymbol{B}}$ may be continuous everywhere, it may happen that div $\tilde{\boldsymbol{B}}$ is non-zero, because it does not coincide with the true value of $\boldsymbol{B}$. It has been shown that the use of such an unphysical field gives rise to spurious forces on the charges. This can be avoided by interpolating the vector potential $\boldsymbol{A}$ instead and taking $\boldsymbol{B} = \text{curl } \boldsymbol{A}$. If a two-dimensional cubic spline is used in the interpolation, the derivatives of $\boldsymbol{A}$, and hence $\boldsymbol{B}$, will be continuous, and since the field is obtained as an exact curl of a function its divergence vanishes exactly.

1.4 Parametrisation of a Function using Fourier Coefficients

A function $y(x)$ which is periodic in L (i.e. $y(x) = y(x + L)$) may be expanded as a Fourier series

$$y(x) = \frac{A_0}{2} + \sum_{n=1}^{\infty} \left[A_n \cos\left(\frac{2\pi nx}{L}\right) + B_n \sin\left(\frac{2\pi nx}{L}\right) \right] \tag{1.12}$$

where, because of the orthogonality of sine and cosine functions over the range L, the coefficients A_n and B_n are given by the expressions

$$\begin{aligned} A_n &= \frac{2}{L} \int_0^L y(x) \cos\left(\frac{2\pi nx}{L}\right) \mathrm{d}x \\ B_n &= \frac{2}{L} \int_0^L y(x) \sin\left(\frac{2\pi nx}{L}\right) \mathrm{d}x. \end{aligned} \tag{1.13}$$

A typical situation is that experimental values of the function $y(x)$ have been obtained and we wish to obtain the corresponding parameters A_n, B_n for values of $n < N$. If appreciably more than $2N + 1$ values of $y(x)$ are known, but for scattered values of x, then least squares fitting directly to equation (1.12) may be appropriate. If, on the other hand, $y(x)$ is known for $2N + 1$ or more evenly spaced values of x, then equation (1.13) may be evaluated using standard numerical integration procedures. In particular, if $2N + 1$ values of $y(x)$ are known, then equations (1.12) and (1.13) define a finite form of the Fourier transformation which relates two sets of $2N + 1$ quantities. There will, of course, be differences between the properties of this transformation and the corresponding integral transform which has $L \to \infty$ and $N \to \infty$. In this section we shall concentrate on the finite transformation approach, as this involves no loss of the information provided by the input values of $y(x)$. The representation of a continuous function by a finite set of parameters will be discussed in §1.5.

The finite difference (trapezoidal rule) expressions corresponding to equations (1.12) and (1.13) are (with $L = M\Delta x$, $M = 2N$, $y_m = y(m\Delta x)$)

$$f(x) = \frac{A_0}{2} + \sum_{n=1}^{N} \left[A_n \cos\left(\frac{2\pi nx}{L}\right) + B_n \sin\left(\frac{2\pi nx}{L}\right)\right] \tag{1.14}$$

$$\begin{aligned} A_n &= \frac{2}{M} \sum_{m=0}^{M} y_m \cos\left(\frac{2\pi nm}{M}\right) - \frac{1}{M}(y_0 + y_M) \qquad n = 0, \ldots, N \\ B_n &= \frac{2}{M} \sum_{m=0}^{M} y_m \sin\left(\frac{2\pi nm}{M}\right) \qquad n = 1, \ldots, N. \end{aligned} \tag{1.15}$$

The main considerations in evaluating the sums in equation (1.15) are speed and accuracy. A test on any calculator or microcomputer will soon convince the reader that preprogrammed routines for evaluating cosine and sine functions are far from instantaneous, although they may be quick enough when relatively few evaluations are required. To improve on such routines we seek to take advantage of the special properties of periodic functions.

One useful procedure is to use the trigonometric addition theorems to provide *exact* recurrence relations to generate the sine and cosine functions stepwise over their whole range. We have, for example,

$$\cos\left(\frac{2\pi n(m+1)}{M}\right) = 2\cos\left(\frac{2\pi nm}{M}\right)\cos\left(\frac{2\pi n}{M}\right) - \cos\left(\frac{2\pi n(m-1)}{M}\right). \tag{1.16}$$

This equation, and a similar one for the sine function, provides a relatively rapid generation of the functions required in a particular sum from a single calculated value, $\cos(2\pi n/M)$. The accuracy of such procedures depends, of course, on the number of digits carried by the computer, but it is easy to make suitable checks (e.g. by occasional use of the 'built-in' functions).

Another method would be to generate complete matrices of the functions $\cos(2\pi nm/M)$ and $\sin(2\pi nm/M)$ and express equation (1.15) in terms of matrix multiplication. This is most suitable for cases where a succession of functions $y(x)$ has to be transformed, but is likely to be too demanding on storage space to be appropriate for a microcomputer.

A far more systematic method, called *fast Fourier transformation*, has been devised to take full advantage of relationships that exist between cosine and sine functions with arguments $(2\pi nm/M)$. This method is based on carrying out a series of intermediate transformations, each of which uses matrices of much smaller dimension than the $M \times M$ matrix of coefficients required to evaluate equation (1.15).

The method of analysis is most simply explained by using complex parameters and combining the pairs of equations (1.13) or (1.15). Writing $C_n = (2/M)(A_n + iB_n)$, and redefining A_n to omit the constant correction term, we have the simplified expression

$$C_n = \sum_{m=1}^{M} y_m \exp i\left(\frac{2\pi nm}{M}\right) \qquad (n < N = m/2).$$

Dividing this into sums over alternate values of m we have

$$C_n = \sum_{m=1}^{M/2} \left[y_{2m} \exp \mathrm{i} \left(\frac{4\pi nm}{M} \right) + y_{2m-1} \exp \mathrm{i} \left(\frac{2\pi n}{M} (2m-1) \right) \right]$$

$$= D_n + \exp \left(-\mathrm{i} \frac{2\pi n}{M} \right) E_n \tag{1.17}$$

where D_n and E_n are thus Fourier coefficients corresponding to half the data.

It is now possible to relate the values of C_n for $n > N/2$ to those with $n < N/2$. We write, putting $(N - n)$ for n in equation (1.17),

$$C_{(N-n)} = \sum_{m=1}^{M/2} \left[y_{2m} \exp \mathrm{i} \left(\frac{4\pi(N-n)m}{M} \right) + y_{2m-1} \exp \mathrm{i} \left(\frac{2\pi(N-n)(2m-1)}{M} \right) \right]$$

$$= \sum_{m=1}^{M/2} \left[y_{2m} \exp \mathrm{i} \left(-\frac{4\pi nm}{M} \right) - y_{2m-1} \exp \mathrm{i} \left(-\frac{2\pi n(2m-1)}{M} \right) \right]$$

$$= D_n^* - \exp \left(\mathrm{i} \frac{2\pi n}{M} \right) E_n^* \tag{1.18}$$

where stars denote complex conjugation. Hence, all the C_n values can now be deduced from the calculated values of D_n and E_n for $n < N/2$, each of which only requires one quarter the number of evaluations of cosine and sine functions. This reduces the computation by a factor of approximately one half.

The process described above can, in practice, be used several times so that we can express the C_n as linear combinations of 2, 4, 8, 16 etc—partial Fourier transforms. The factor two advantage at each doubling will, however, gradually be eroded by the increasing number of terms in the equations corresponding to (1.17) and (1.18). It is also necessary, of course, to have a factor 2^r in M if we wish to carry out this process r times. Nevertheless, the computer time saved by the fast Fourier transform method is very real, as the (maximal) time saving factor of two for each doubling indicates.

Although the algebra is a little more tedious, exactly the same saving of effort can be shown to occur for *real* Fourier transforms.

1.5 Least Squares Fitting

Linear least squares fitting, in which the fitted expression is linear in the parameters, may be regarded as a special case of least squares methods for general (i.e. non-linear) expressions. We shall find, however, that its special features make it worthy of detailed study in its own right. Perhaps the most important of these features is that we are able to obtain an explicit formula for the fitted parameters. Unfortunately, real life problems frequently cannot be reduced to linear relationships with any degree of accuracy. Several situations of this nature will be discussed in Chapter 2, which also contains some general remarks on the method of carrying out non-linear least squares fits.

It is normally assumed that the input data have the form $(x_i, y_i \pm \sigma_i)$, $i = 1, \ldots, N$. With each 'observation' y_i there is associated an uncertainty σ_i, which has been estimated in relation to the experimental procedure. The x_i values distinguish different observations and are presumed to be known exactly.

Linear least squares fitting is based on the presumption that a set of functions $f_\alpha(x_i)$ can be found such that the expression for y is linear in a set of M $(<N)$ parameters a_α, i.e.

$$y(x) = \sum_{\alpha=1}^{M} f_\alpha(x) a_\alpha. \tag{1.19}$$

Note that the only difference between this equation and (1.1) is that the number M of parameters a_α is less than the number of data values y_i. As we demonstrate below, explicit formulae can be found for the a_α values if we minimise the sum of squares of deviations between the data values y_i and the $y(x_i)$ given by the fitting formula (1.19).

Let the deviations between the data and the function y be written

$$h_i = y_i - y(x_i) = y_i - \sum_\alpha f_\alpha(x_i) a_\alpha. \tag{1.20}$$

The least squares fit is obtained by minimising the weighted sum of squares of deviations

$$D = \sum_{i=1}^{N} \frac{h_i^2}{\sigma_i^2} \tag{1.21}$$

with respect to the parameters a_α. Hence, we require $\partial D/\partial a_\alpha = 0$ for all α. This gives M linear equations in the M unknowns a_α:

$$2\sum_{i=1}^{N} \frac{1}{\sigma_i^2} h_i \frac{\partial h_i}{\partial a_\alpha} = 2\sum_{i=1}^{N} \frac{1}{\sigma_i^2}\left[y_i - \sum_\beta f_\beta(x_i) a_\beta\right][-f_\alpha(x_i)] = 0.$$

A little algebra enables us to express these equations in the matrix form

$$(\mathbf{A}^\mathrm{T}\mathbf{W}\mathbf{A})\mathbf{a} = \mathbf{A}^\mathrm{T}\mathbf{W}\mathbf{y} \tag{1.22}$$

where the weight matrix $\mathbf{W}$ is given by

$$\mathbf{W} = \begin{bmatrix} \frac{1}{\sigma_1^2} & & & \\ & \frac{1}{\sigma_2^2} & & \\ & & \ddots & \\ & & & \frac{1}{\sigma_N^2} \end{bmatrix} \tag{1.23}$$

and

$$\mathbf{A} = \begin{bmatrix} f_1(x_1)\, f_2(x_1) \ldots\ldots\ldots\ldots\ldots\ldots . f_M(x_1) \\ f_1(x_2) \ldots\ldots\ldots\ldots\ldots\ldots\ldots\ldots . f_M(x_2) \\ \ldots\ldots\ldots\ldots\ldots\ldots\ldots\ldots\ldots\ldots \\ f_1(x_N) \ldots\ldots\ldots\ldots\ldots\ldots\ldots\ldots f_M(x_N) \end{bmatrix} \tag{1.24}$$

The square matrix $\mathbf{A}^{\mathrm{T}}\mathbf{W}\mathbf{A}$ may, in general, be inverted to obtain explicit formulae for the parameter estimates $\hat{\boldsymbol{a}}$. These may be written collectively as

$$\hat{\boldsymbol{a}} = (\mathbf{A}^{\mathrm{T}}\mathbf{W}\mathbf{A})^{-1}\,\mathbf{A}^{\mathrm{T}}\mathbf{W}\mathbf{y} \tag{1.25}$$

which generalises equation (1.3).

The diagonal form of $\mathbf{W}$ is based on the assumption that the errors in the data values y_i are uncorrelated. We can sometimes go further than this and assume that $\sigma_i = \sigma$ (all i) so that $\mathbf{W} = (1/\sigma^2)\mathbf{I}_N$, where $\mathbf{I}_N$ is the unit matrix of dimension N. It cannot be assumed, however, that the fitted parameters a_α are uncorrelated, as we shall see below.

The variance–covariance matrix of a data vector z is the matrix $\mathbf{Z}$ with components

$$Z_{ij} = \langle (z_i - \langle z_i \rangle)(z_j - \langle z_j \rangle) \rangle$$

where the expectation values, denoted $\langle\rangle$, refer to notional averages taken over many sets of data. The off-diagonal elements correspond to covariances (e.g. see §A1 of the Appendix). It follows that a linear transform of the data (say $\mathbf{B}z$, where $\mathbf{B}$ is a constant matrix) has a variance–covariance matrix with components

$$Z'_{ij} = \left\langle \sum_k B_{ik}(z_k - \langle z_k \rangle) \sum_m B_{jm}(z_m - \langle z_m \rangle) \right\rangle$$

or, in matrix notation, $\mathbf{Z}' = \mathbf{B}\mathbf{Z}\mathbf{B}^{\mathrm{T}}$. If we now identify $z = y$, $\mathbf{Z} = \mathbf{W}^{-1}$ and let $\mathbf{B} = (\mathbf{A}^{\mathrm{T}}\mathbf{W}\mathbf{A})^{-1}\mathbf{A}^{\mathrm{T}}\mathbf{W}$ correspond to equation (1.25), we obtain the following expression for the variance–covariance matrix of the $\hat{a}_\alpha$ values:

$$\begin{aligned} (\mathbf{W}')^{-1} &= (\mathbf{A}^{\mathrm{T}}\mathbf{W}\mathbf{A})^{-1}(\mathbf{A}^{\mathrm{T}}\mathbf{W}\mathbf{A})(\mathbf{A}^{\mathrm{T}}\mathbf{W}\mathbf{A})^{-1} \\ &= (\mathbf{A}^{\mathrm{T}}\mathbf{W}\mathbf{A})^{-1}. \end{aligned} \tag{1.26}$$

In the particular case $\mathbf{W} = (1/\sigma^2)\mathbf{I}_N$, $\mathbf{W}' = (1/\sigma^2)\,\mathbf{A}^{\mathrm{T}}\mathbf{A}$ which, as we mentioned above, is not necessarily diagonal, the off-diagonal elements in $(\mathbf{W}')^{-1}$ are interpreted as the covariances between different a_α.

Correlation (i.e. non-zero covariance) between the a_α values can be removed entirely if the functions used in the fitting satisfy orthogonality relations

$$\sum_i f_\alpha(x_i) f_\beta(x_i) \sigma_i^{-2} = 0 \qquad \text{for all } \alpha \neq \beta \tag{1.27}$$

as this makes the matrix $\mathbf{A}^{\mathrm{T}}\mathbf{A}$ diagonal. In most cases, however, it will involve a considerable amount of calculation to modify fitting functions so as to satisfy this criterion. Besides this, the choice of functions (and their corresponding parameters) will then depend on the particular set of data being analysed. It is usually preferable, therefore, to employ functions taken from an orthonormal (orthogonal and normalised) set, defined over the continuous range (R say) in which the x_i values occur. That is to say, we choose functions $f_\alpha(x)$ which satisfy

$$\int_R \frac{f_\alpha(x) f_\beta(x)}{\sigma^2(x)} \, \mathrm{d}x = \delta_{\alpha\beta} \begin{cases} = 0 \text{ if } \alpha \neq \beta \\ = 1 \text{ if } \alpha = \beta. \end{cases} \tag{1.28}$$

This is, of course, not equivalent to the previous equation, but it should be approximately the same if the x_i are spread fairly uniformly over the range R.

As an example, we consider fitting to the polynomial functions $f_n(x) = x^{n-1}$. In this case it is better to use the Legendre functions $P_n(x)$ if the x_i values span a finite range. These functions are orthonormal over the range $-1 < x < 1$, i.e.

$$\int_{-1}^{1} P_n(x) P_m(x) \, \mathrm{d}x = \frac{2}{2n+1} \delta_{nm} \tag{1.29}$$

where $\delta_{nm} = 1$ if $n = m$ and zero otherwise. The low order polynomials are

$$P_0 = 1$$

$$P_1 = x$$

$$P_2 = \tfrac{1}{2}(3x^2 - 1)$$

$$P_3 = \tfrac{1}{2}(5x^3 - 3x)$$

$$P_4 = \tfrac{1}{8}(35x^4 - 30x^2 + 3)$$

$$P_5 = \tfrac{1}{8}(63x^5 - 70x^3 + 15x)$$

$$P_6 = \tfrac{1}{16}(231x^6 - 315x^4 + 105x^2 - 5).$$

Higher order polynomials can be generated using the orthonormality criterion (equation 1.29), or more directly from the recurrence relation

$$P_{n+1}(x) = [1/(n+1)][(2n+1)xP_n(x) - nP_{n-1}(x)].$$

They can be adapted to any finite range by a suitable change of variable using the expression $x' = ax + b$.

If the data are taken at $(N+1)$ uniformly spaced intervals in the sequence $x = 0, 1, 2, \ldots, N$, it is appropriate to use the Chebyshev polynomials $C_n(x)$, which can be generated recursively as follows

$$C_0(x) = 1$$

$$C_1(x) = 1 - (2x/N)$$

$$(n+1)(N-n)C_{n+1} = (2n+1)(N-2x)C_n - n(N+n+1)C_{n-1}.$$

These functions can be shown to satisfy equation (1.27) if $\sigma_i = \sigma$ (all i).

There are several reasons for using orthogonal functions in least squares fits. One of these is that it can be shown that the matrix $\mathbf{A}^T\mathbf{A}$ may become ill-conditioned when non-orthogonal fitting functions are used. This is certainly the case for the polynomial functions (for example, see Hudson (1964)). A second reason is a preference for uncorrelated parameters. They have two advantages: additional parameters can be added without changing the values of those that have already been fitted and the parameters have minimum errors as compared with an equivalent set of non-orthogonal fitting functions.

Many reported applications of the least squares method do not employ criteria for assessing their success. Nevertheless, it is important to check that the variance of the parameters is reasonably related to the variance of the data. If the former is much smaller than the latter we must be fitting the random scatter; if it is much larger we are getting a poor fit. This may be due to the use of an inappropriate type, or an insufficient number, of fitting functions. Figure 1.5 illustrates these remarks. A detailed discussion of this problem may be found in Larimore and Mehra (1985).

If we assume that the individual observations y_i are samples from a normal distribution, then the sum of the squared deviations $D = \Sigma_i h_i^2/\sigma_i^2$ is distributed according to the χ^2 distribution, and the fitted value of D may be subjected to the standard χ^2 'goodness of fit' test, discussed in §A3 of the

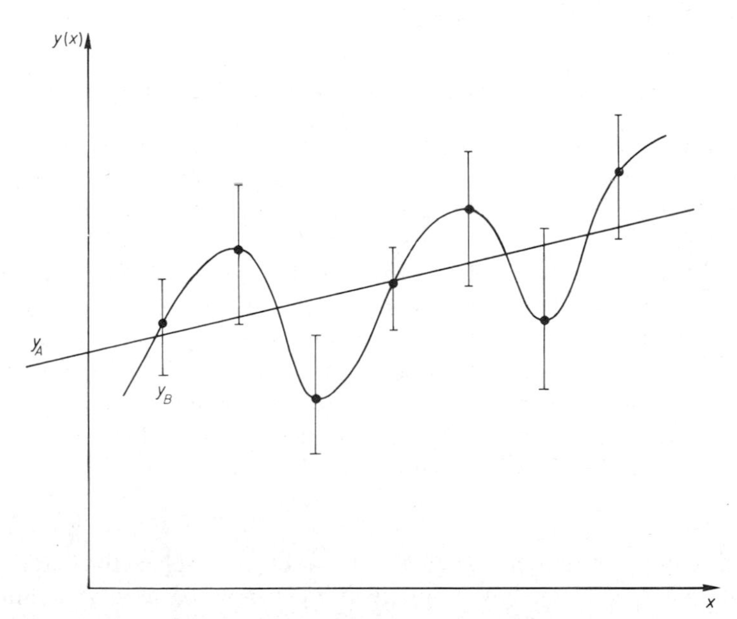

Figure 1.5 Data with errors bars, showing a straight line fit (y_A) and a spurious fit (y_B) of the random scatter to a polynomial function.

Appendix. In this application the relevant number of degrees of freedom is equal to the number of data minus the number of fitted parameters.

Using table A3.1, we see that when correctly fitting 10 data points with 6 parameters (i.e. with $v = 10 - 6 = 4$), a value of $D\,(=T)$ greater than 3.3 (corresponding to $u = -0.25$) would be expected in about 50% of cases, while a value greater than 9.5 (corresponding to $u = 1.95$) would be expected in only about 5% of cases. In this case, a value of $D = 3.3$ thus provides evidence for the correctness of the model, while $D = 9.5$ is evidence that the model is not working. On the other hand, a value of $D = 0.7$, corresponding to a probability of 0.95 that D could be larger, suggests that the fit is 'too good', that is to say, it looks as if the data errors are being fitted. Such a situation can arise when the structure of the model is such that the equations relating data to parameters are ill-conditioned, so that the number of effective degrees of freedom is really lower than calculated.

The main computational problems to be solved in programming equation 1.25 are matrix multiplication and matrix inversion. It is thus preferable to employ a computer which accepts these operations as direct commands or has appropriate subroutines available. At the time of writing, however, the cheaper range of personal microcomputers does not provide a sufficiently extended BASIC interpreter for this. Such functions are also not available in PASCAL compilers. Hence, it may be found preferable to use the linear least squares fitting program in a standard package (for example, see Nash (1985)).

Generally speaking, direct procedures for solving the simultaneous equations

$$\mathbf{B}\boldsymbol{a} = \boldsymbol{z} \qquad \text{where } \mathbf{B} = \mathbf{A}^{\mathrm{T}}\mathbf{W}\mathbf{A} \text{ and } \boldsymbol{z} = \mathbf{A}^{\mathrm{T}}\mathbf{W}\boldsymbol{y}$$

for the quantities a_α will work faster than procedures which obtain the inverse of the matrix $\mathbf{B}$ explicitly. If many sets of input data $\boldsymbol{z}$ are to be parametrised for a given $\mathbf{B}$, there may nevertheless be some advantage in obtaining $\mathbf{B}^{-1}$.

In the absence of any standard procedure for solving simultaneous equations, the reader will find it most convenient to construct a program using Gaussian elimination. The first step is to eliminate a parameter (a_1 say) from all equations except one by subtracting a suitable multiple of the chosen equation from all the others. The parameter a_2 is similarly eliminated from $(m-2)$ equations, a_3 from $(m-3)$ equations and so on. Finally, the set of simultaneous equations will be reduced to the form

$$\mathbf{B}'\boldsymbol{a} = \boldsymbol{z}'$$

where $\mathbf{B}'$ can be written in triangular form with all the elements (say) below the diagonal zero. A systematic procedure can now be used to obtain the complete solution, starting with the final equation which has a single unknown and working back through the equations in the reverse order to that in which they were obtained.

Gaussian elimination is relatively easy to program, especially if no safeguards have to be built in to avoid small or vanishing coefficients of the parameters a_α. A simple way of overcoming this problem is to use an interactive program so that the operator chooses which variable and which equation are to be eliminated at each stage. This both simplifies program writing and makes it easy to check that it is proceeding correctly.

Exercises

Exercises 1.1 to 1.4 are purely algebraic, the remaining exercises require a least squares fitting program.

1.1 Fill in the algebraic steps leading to equation (1.22).

1.2 Check that the polynomial expression for P_5 is correct and find expressions for P_7 and P_8.

1.3 Find the transformed functions P_n $(n = 1, \ldots, 5)$ which are orthonormal over the range $0 < x < 4$.

1.4 By adapting their range, find the relation between the low order Chebyshev and Legendre polynomials.

1.5 Use the following sets of data to carry out least squares fits to a polynomial function. In particular, you should check the stated advantages of using various types of orthogonal fitting functions. You will find it useful to incorporate a range transformation routine into your program.

Trial Data (x_i, y_i)
(a) (5, 0.1), (4.5, −205.3), (3, −161.6), (2, −48.3), (1.5, −17.5), (1, −3.8), (0.5, −0.3), (0, 0.1), (−1, −5.8), (−1.5, −33.1); $\sigma_i = \sigma = 0.3$.
(b) (0, 0.12), (0.5, 1.27), (1, 3.07), (1.5, 13.66), (2, 65.89), (−0.5, 1.77), (−1, −3.11); $\sigma_i = \sigma = 0.05$.

References

Cheney W and Kincaid D 1980 *Numerical Mathematics and Computing* (Monterey, CA: Brooks-Cole)

Hudson D J 1964 *Statistics Lectures II: Maximum Likelihood and Least Squares Theory* (Geneva: CERN 64–18)

Larimore W E and Mehra R K 1985 The Problem of Overfitting Data *Byte* **10** (October 1985) 167–80

Nash J C 1985 Scientific Applications Software *Byte* (December 1985) **10** 145–50

Rossi B and Greisen K 1941 *Rev. Mod. Phys.* **13** 240–309

Strong J and Vanasse G A 1959 *J. Opt. Soc. Am.* **49** 844–50

Stroud A H 1974 *Numerical Quadrature and Solution of Ordinary Differential Equations* (New York: Springer)

Chapter 2

Applications of Least Squares Fitting

2.1 Introduction

Least squares fits may be carried out for two distinct purposes. In the first of these they are used simply to fit a smooth curve through many data points, often with the idea of making comparisons with a theoretically determined curve. It is often only necessary to obtain an unambiguous form of the curve to allow for interpolation and extrapolation, as specific fitting functions will not necessarily be suggested by physical considerations. In this case, the values obtained for the fitted parameters have no intrinsic interest. The examples given at the end of Chapter 1 were treated from this point of view.

In the second type of application the fitting procedure is concerned with testing a model and the determination of the best possible values of physically significant parameters. The fitting functions which characterise the model are determined by the physical processes in a general way which leaves the values of the fitting parameters undetermined. Fitted parameter values may then be compared with values calculated from first principles, but such calculations will inevitably involve assumptions and approximations in addition to those which characterise the model used in the least squares fit. The four projects described in this chapter are all of this type, so it will be necessary to go into sufficient detail in each case to characterise the physical model.

An important aspect of any model of physical processes is the set of *minimal hypotheses* which is necessary to establish the form of the mathematical expressions used in fitting the parameters. This concept has been discussed in some detail by Newman (1978). If the set of minimal hypotheses can be identified explicitly, then tests of the model are equivalent to tests of the hypotheses. However, it is not uncommon in physics to overlook the possibility that weaker hypotheses could lead to the same model. This can, of course, lead us badly astray in the interpretation of experimental results. Hence, great care must be taken to establish the minimal hypotheses which characterise models.

In some cases alternative, equally successful, models may exist with apparently inconsistent minimal hypotheses. Such problems can only be resolved by extending the range of applicability of the model so that other types of experimental data can be used in testing its validity. An example of this situation is discussed in relation to Project 2C.

2.2 Non-linear Least Squares Fitting

In some cases, an obvious generalisation of a linear parametrised model suggests the introduction of terms which are non-linear in the parameters. The models used in both Projects 2A and 2C can be generalised in this way. In other cases it is not possible even to begin with a linear model. It is therefore important to find efficient fitting procedures even when the fitting function is not linear in the parameters. Unless a transformation exists which converts the non-linear problem into a linear one, recourse must be had to iterative methods which essentially break down the non-linear problem into a succession of linear problems for small changes in the parameters. The mathematical details of this are discussed below so that the reader may adapt a linear least squares program for use in the non-linear case. Alternatively, it may be found preferable to use an optimisation program from a standard package, in which case the remainder of this section can be omitted.

Instead of equation (1.19), let us consider a general function of x and parameters $\boldsymbol{\theta} = \{\theta_\alpha\}$:

$$y = f(x, \boldsymbol{\theta}). \tag{2.1}$$

We shall exploit the fact that, to first order, changes in the value of y due to small errors in the estimated values $\hat{\boldsymbol{\theta}}$ of the parameters $\boldsymbol{\theta}$ can be expressed as

$$\delta y_i = f(x_i, \boldsymbol{\theta}) - f(x_i, \hat{\boldsymbol{\theta}}) = \sum_\alpha \left(\frac{\partial f}{\partial \theta_\alpha}\right)_{\boldsymbol{\theta} - \hat{\boldsymbol{\theta}}} \delta\theta_\alpha + \mathrm{O}(\delta\theta_\alpha^2) \tag{2.2}$$

which is analogous to (1.19) with the replacement of a_α by $\delta\theta_\alpha$. Given a set of values (x_i, y_i), equation (2.2) provides the means to improve the estimate of $\boldsymbol{\theta}$, given an initial set of estimates $\boldsymbol{\theta} = \hat{\boldsymbol{\theta}}$. Such a procedure can be used iteratively, leading eventually (if the equations are 'well behaved') to a stable minimisation of errors.

At each iteration the quantity to be minimised is given by equation (1.21) with

$$h_i = y_i - f(x_i, \boldsymbol{\theta}) = y_i - \left[f(x_i, \hat{\boldsymbol{\theta}}) + \sum_\alpha \left(\frac{\partial f_i}{\partial \theta_\alpha}\right)_{\boldsymbol{\theta} = \hat{\boldsymbol{\theta}}} \delta\theta_\alpha\right]. \tag{2.3}$$

Following the standard least squares procedure for the $\delta\theta_\alpha$ we obtain an equation analogous to equation (1.25) where the matrix $\mathbf{A}$ has elements $\partial f_i/\partial\theta_\alpha$, $\hat{a} \rightarrow \delta\boldsymbol{\theta}$ and $\boldsymbol{y} \rightarrow \delta\boldsymbol{y} - \boldsymbol{f}(\hat{\boldsymbol{\theta}})$. Use of these equations will be referred to as *linear iteration*.

When equation (2.3) is substituted into (1.21), it produces a paraboloid which has the same gradients $\partial D/\partial \theta_\alpha$ at the point where it touches the D surface in parameter space at $\hat{\boldsymbol{\theta}}$. The value of $\boldsymbol{\theta}$ at the (single) minimum of this paraboloid then provides the next estimate of the parameters $\boldsymbol{\theta} = \hat{\boldsymbol{\theta}} + \delta\boldsymbol{\theta}$. The reason that we can determine the minimum from a knowledge of the gradients alone is due to the assumption that the second-order differentials of $f(x, \boldsymbol{\theta})$ with respect to the θ_α can be neglected. In fact, it would be more appropriate to replace (2.3) with

$$h_i = y_i - \left[f(x_i, \hat{\boldsymbol{\theta}}) + \sum_\alpha \left(\frac{\partial f_i}{\partial \theta_\alpha} \right)_{\boldsymbol{\theta} = \hat{\boldsymbol{\theta}}} \delta\theta_\alpha + \sum_{\alpha\beta} \left(\frac{\partial^2 f_i}{\partial \theta_\alpha \, \partial \theta_\beta} \right)_{\boldsymbol{\theta} = \hat{\boldsymbol{\theta}}} \delta\theta_\alpha \, \delta\theta_\beta \right]. \tag{2.4}$$

This can be substituted into equation (1.21) and all terms quadratic in $\delta\boldsymbol{\theta}$ preserved in the minimisation of D. Taking all $\sigma_i = \sigma$ (for simplicity) the fitting equations then become

$$\sum_{\beta i} \left[\left(\frac{\partial f_i}{\partial \theta_\alpha} \right)_{\hat{\boldsymbol{\theta}}} \left(\frac{\partial f_i}{\partial \theta_\beta} \right)_{\hat{\boldsymbol{\theta}}} - 2(y_i - f_i(\hat{\boldsymbol{\theta}})) \frac{\partial^2 f_i}{\partial \theta_\alpha \, \partial \theta_\beta} \right] \delta\theta_\beta = \sum_i \left(\frac{\partial f_i}{\partial \theta_\alpha} \right)_{\hat{\boldsymbol{\theta}}} (y_i - f_i(\hat{\boldsymbol{\theta}})). \tag{2.5}$$

These equations should provide a more accurate estimate of the minimum than the usual equations used in non-linear least squares fitting which omit second derivatives.

There may be difficulties with the linear iteration process, especially if $\delta\boldsymbol{y}$ is similar in magnitude to $\boldsymbol{y}$. This can result in very large changes $\delta\boldsymbol{\theta}$ which may not even result in an overall reduction of $D = \Sigma_i h_i^2/\sigma_i^2$. Jones (1970) has suggested a strategy based on (i) testing whether linear iteration reduces D and, if not, (ii) using an iteration procedure which mixes linear iteration and the steepest descent method.

The steepest descent method consists of changing the parameters in a direction corresponding to the negative gradient of D in parameter space. This corresponds to taking

$$\delta\theta_\beta = \lambda^2 \sum_i (y_i - f(x_i, \hat{\boldsymbol{\theta}})) \frac{\partial f_i}{\partial \theta_\beta} \tag{2.6}$$

where λ determines the magnitude of the changes in θ_β and must be chosen small enough to ensure that we do not overshoot the minimum. In the method put forward by Jones (1970), a systematic approach is used to determine the value of λ^2 at each step of the iteration and the changes $\delta\boldsymbol{\theta}$ specified by (2.6) are added to those given by least squares fitting.

In general terms, we expect the steepest descent approach to be most appropriate when $\hat{\boldsymbol{\theta}}$ is some way from the minimum (so that gradients are large) and linear iteration to be best when $\hat{\boldsymbol{\theta}}$ is close to the minimum. It is often possible to get good initial estimates of parameters either by comparison with similar systems, or by careful examination of the data prior to fitting. If this is possible, then linear iteration will usually be sufficient.

Quite apart from the difficulties of actually arriving at a minimum, we must remember that non-linear functions may produce a very complicated surface $D(\boldsymbol{\theta})$ in parameter space, with many minima (and maxima). If there are several minima, we usually have to resort to the physical interpretation of the parameters in order to determine the 'correct' one. This will not necessarily correspond to the smallest value of D. Generally speaking, any non-linear least squares fit should be tried with several different starting values of the θ_α value and large $\delta\theta_\alpha$ should be avoided by scaling down or using steepest descent methods. Overall, non-linear least squares fitting can best be carried out using an interactive program, providing for both linear iteration and steepest descent methods, so that the operator can get some feeling for the sort of surface, $D(\theta_\alpha)$, he or she is dealing with and choose an appropriate strategy. In some cases, it will be best to use the steepest descent method for initial iterations and then to use the linear iteration method to refine the solution.

A discussion of the mathematical aspects of dealing with poorly conditioned minimisatiom equations can be found in the book *Solving Least Square Problems* by Lawson and Hanson (1974).

PROJECT 2A: SPIN–LATTICE RELAXATION

The basic idea of spin–lattice relaxation is that the energy of isolated electronic systems, such as paramagnetic ions in crystals, can leak away into the crystal lattice in the form of vibrational energy. This is referred to as *spin–*lattice relaxation because it is usual to label the low lying many-electron states of isolated electronic systems by 'effective' spin quantum numbers. According to quantum mechanics, relaxation is a statistical process in which each system in the assembly of isolated systems has a definite probability of losing a quantum of energy. Several distinct mechanisms are possible, and it has been shown by Orbach (1961a,b, 1962) and others that each mechanism has a characteristic temperature dependence which is independent of the strength of the coupling between the electronic systems and the lattice. Each of these mechanisms operates independently, so that the total rate of loss of energy (proportional to the inverse of the relaxation time τ) can be expressed as a sum of the rates due to each mechanism. Given the experimental relaxation rate as a function of temperature, we can therefore fit it to a sum of functions of the temperature T by varying the coefficients which represent the coupling strengths of the different processes.

Figure 2.1 shows the three types of relaxation process that are usually taken into account in a simple two-level system. In the thermodynamic derivation of the temperature dependence of the processes represented in the figure it is found necessary to distinguish between two types of physical systems. The two levels may be non-degenerate in the absence of a magnetic field or they may be strictly degenerate so that a magnetic field has to be applied to separate the levels. In the latter case they are referred to as a 'Kramers degenerate pair'.

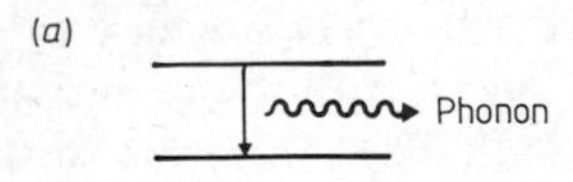

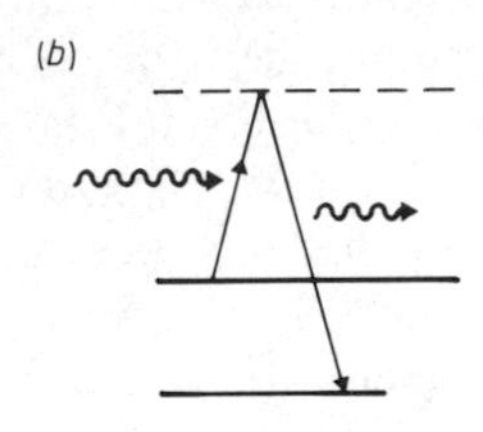

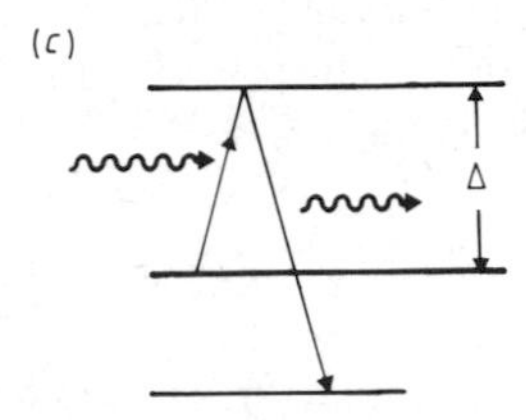

Figure 2.1 (*a*) Direct process, (i), in which the electronic energy is absorbed by the production of a phonon. (b) Raman process, (ii), in which the electronic energy increases the energy of a phonon. The electronic excited state is virtual. (c) Orbach process, (iii), corresponding to the absorption of a phonon which increases the electronic energy followed by the emission of a phonon of higher energy. Here the electronic excited state is real.

In the case of non-Kramers states the fitting function may be written (Orbach 1961a,b, 1962)

$$\frac{1}{\tau} = \overset{\text{(i)}}{aT} + \overset{\text{(ii)}}{bT^7} + \overset{\text{(iii)}}{c\exp(-\Delta/k_BT)} \tag{2.7}$$

in which the labels above the terms correspond to the processes shown in figure 2.1. The terms a, b and c are the parameters to be fitted, k_B is Boltzmann's constant and Δ is the energy of the excited state in process (iii). In the case of Kramers states we have a rather different formula (Orbach 1961a,b, 1962), namely

$$\frac{1}{\tau} = \overset{\text{(i)}}{a'H^4T} + \overset{\text{(ii)}}{b'T^9} + \overset{\text{(ii)}}{b''H^2T^7} + \overset{\text{(iii)}}{c'\exp(-\Delta/k_BT)} \tag{2.8}$$

where H represents the strength of the magnetic field.

The excited state energy Δ may be known from other experiments. If not, it is often possible to determine its value by fitting to the relaxation rates. Sometimes, in fact, this provides a convenient way of finding the position of low lying states. In order to fit Δ it is necessary to introduce a procedure for making *non-linear* least squares fits. The simplest way of doing this is to use an iterative method in which

(i) Δ is held constant for a normal linear fit, and then
(ii) only the constants c and Δ of the last term are fitted to $\log(1/\tau)$, then (i) is repeated until convergence is achieved.

In practice, however, it often occurs that the $\exp(-\Delta/k_B T)$ is dominant over some range of T, so that only the last term need be retained in the fit over this restricted range.

Spin–lattice relaxation data are not customarily published in tabular form, so that it will be necessary for the reader to take data values from graphs in order to carry out numerical work on this example. Abragam and Bleaney (1970) give two examples as well as providing more background material on

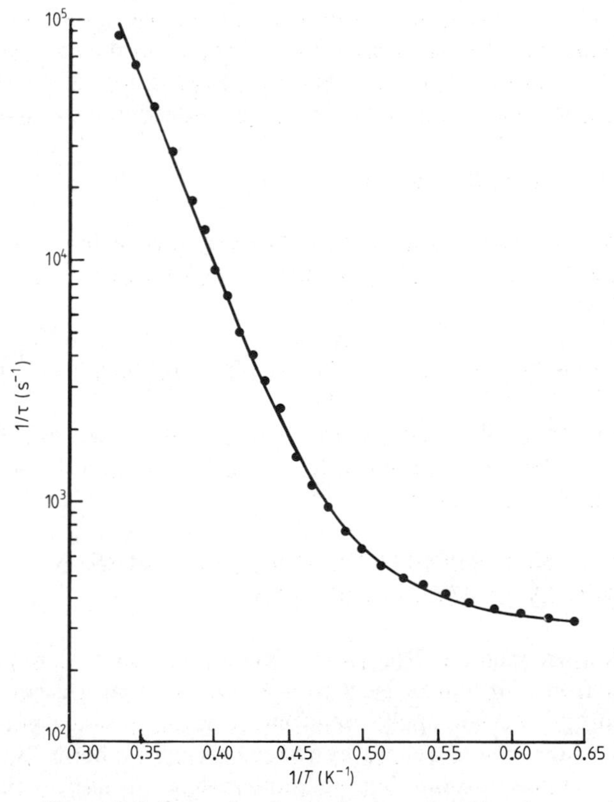

Figure 2.2 Example of spin–lattice relaxation data.

the model. An example of the type of data fitted (from Fish *et al* (1980)) is shown in figure 2.2. It must be emphasised that it is not necessary to understand anything about the detailed mechanisms of relaxation in order to carry out fitting procedures using equations (2.7) and (2.8). The questions that should be asked are:

Does the published fit agree with yours?
Is the fit good enough to support the model?
How accurately can the parameters be derived?

Finally, we should mention the problem of calculating the values of the fitted parameters from first principles. This is very difficult in this case because of the complicated nature of the coupling between the electronic systems and the lattice. For this reason it is not usual to find anything better than order of magnitude first principles calculations in the literature. It is also difficult to identify precise minimal hypotheses for this model as so many assumptions appear to be necessary about the form of the lattice vibrations. Nevertheless, it is clear that the formulae given above are independent of the details of the coupling mechanism between the electronic states and lattice vibrations and that a very simple model (known as the Debye model) of the lattice vibrations suffices. The question of minimal hypotheses does not appear to have been discussed in the literature. It seems very unlikely, however, that the form of the expressions (2.6)–(2.8) depends sensitively on model assumptions. They may indeed be very general.

Exercise (developing Project 2A)

2.1 In a paper by Kurkin and Tsvetkov (1970) it is stated that the relaxation data for the Er^{3+} ion substituted into $BaWO_4$ can be fitted to the expression

$$\frac{1}{\tau} = 0.88 \times 10^7 \exp\left(-\frac{12.3}{T}\right) + 4.22 \times 10^{10} \exp\left(-\frac{35}{T}\right).$$

What does this tell us about the system? (The paper by Kurkin and Tsvetkov (1970) gives graphical data and fits for several related systems containing the Er^{3+} ion.)

PROJECT 2B: SUPERPOSITION MODEL OF CRYSTAL FIELDS ACTING ON PARAMAGNETIC IONS

It is well known that the electrostatic potential $V(\boldsymbol{r})$ at a point $\boldsymbol{r}$ can be constructed from the sum of electrostatic potentials produced by different charged sources. A crude approximation to the electrostatic potential in an ionic crystal can be constructed by reducing each (spherically symmetric) ionic charge cloud to a point charge and carrying out lattice sums over the point sources to determine the net potential function. We shall first formulate our model in relation to this (traditional) approximation, and then state the minimal hypotheses that lead to the same mathematical structure.

Consider a source-free spherically symmetric region (later to be filled with a paramagnetic ion) embedded in an array of point charges Z_i (see figure 2.3). The potential function may be expanded in terms of the spherical harmonics $Y_{nm}(\theta\phi)$ as (for example, see Jackson 1962)

$$V(\boldsymbol{r}) = \sum_i \frac{Z_i e^2}{|\boldsymbol{r} - \boldsymbol{R}_i|} = \frac{4\pi}{2n+1} \sum_{n,i} \frac{Z_i r^n e^2}{R_i^{n+1}} \sum_m Y^*_{nm}(\theta, \phi) Y_{nm}(\Theta_i, \Phi_i) \tag{2.9}$$

where $\boldsymbol{r} = (r\theta\phi)$ is any point in the spherical region $r < R_i$ (min) and the $\boldsymbol{R}_i = (R_i\Theta_i\Phi_i)$ are the positions of the ionic point charges $Z_i e$, both vectors being measured from a common origin taken to be at the centre of the spherical region. The coefficients of the $\boldsymbol{r}$-dependent factors $r^n Y^*_{nm}(\theta, \phi)$ may be expressed in terms of parameters A_{nm}

$$V(\boldsymbol{r}) = \sum_{nm} A_{nm} r^n Y^*_{nm}(\theta, \phi) \tag{2.10}$$

where, according to (2.9),

$$A_{nm} = \frac{4\pi}{2n+1} \sum_i \frac{Z_i e^2}{R_i^{n+1}} Y_{nm}(\Theta_i, \Phi_i). \tag{2.11}$$

If the point charge at position i is in an axially symmetric position we have $\Theta_i = 0$, giving $Y_{nm}(\Theta_i, \Phi_i) = [4\pi/(2n+1)]^{1/2}\delta_{m,0}$. We may therefore interpret

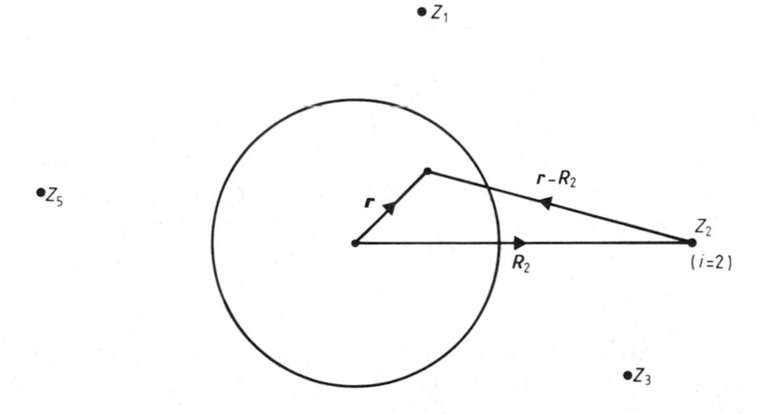

Figure 2.3 Point charge model of the crystal field, showing variables used in equation (2.9).

equation (2.11) as relating the interaction parameters $\bar{B}_n$ for a point charge on the z-axis to the parameters A_{nm} for the complete charge distribution, the coefficients $Y_{nm}(\Theta_i, \Phi_i)$ being determined by the angular position of the point charge:

$$A_{nm} = \sum_i \bar{B}_n(R_i)\left(\frac{4\pi}{2n+1}\right)^{1/2} Y_{nm}(\Theta_i, \Phi_i). \tag{2.12}$$

Although equation (2.12) has been derived in terms of a very simple ionic point charge model, it can in fact be shown to be true for much weaker minimal hypotheses. These characterise the 'superposition model' (Newman 1971, 1978).

(i) The ionic contributions to the total field may be added.
(ii) Each ionic contribution is axially symmetric about the axis joining ion centres.
(iii) Only neighbouring (technically, 'coordinated') ions contribute.

Hypothesis (iii) limits the number of terms in (2.12) and hence limits the number of undetermined parameters $\bar{B}_n(R_i)$. This makes it possible to use the model phenomenologically.

Two further modifications of (2.12) are now made by rearranging (2.9) to ensure that all functions and parameters are real and changing the normalisation to correspond to the frequently employed 'Stevens' notation (Newman 1971). These changes lead to the equation

$$A_n^m\langle r^n\rangle = \sum_i K_{nm}(\Theta_i, \Phi_i)\bar{A}_n(R_i) \tag{2.13}$$

where the real functions $K_{nm}(\Theta_i, \Phi_i)$ are called 'coordination factors' and are related to real combinations of the spherical harmonic functions Y_{nm}. Some of these are given in table 2.1. Equations (2.13) are the linear equations for the $\bar{A}_n(R_i)$ values that will be studied in this project.

The determination of the parameters $A_n^m\langle r^n\rangle$ from the spectroscopically determined energy levels of paramagnetic ions in crystals will be discussed in Chapter 4. Here, we shall just assume that empirical values of these parameters have been established and investigate the use of equation (2.13) to determine the more fundamental single-ion contributions represented by the values of the parameter $\bar{A}_n(R_i)$. In some systems, with fairly low symmetry sites, more empirical parameters $A_n^m\langle r^n\rangle$ are available than the number of different parameters $\bar{A}_n(R_i)$, using all R_i values for a given n. A case in point is the garnet host crystal, for which the paper reprinted in Appendix 2A of this chapter provides many examples.

In practice, the linear model represented by equation (2.13) is tested using $A_n^m\langle r^n\rangle$ values determined from paramagnetic ion spectra. Here the 'test' charges, subject to the total potential, are electrons in the partially filled shell. The spatial distributions of these electrons are characterised by their

Table 2.1 Coordination factors for crystal field parameters in Stevens' normalisation ($A_n^m\langle r^n\rangle$) expressed in terms of spherical polar coordinates. Additional factors are given which adapt these expressions to Wybourne normalisation (B_m^n).

nm	$K_{nm}(\theta\phi)$ for the parameters $A_n^m\langle r^n\rangle$	Factors for B_m^n
40	$\frac{1}{8}(35\cos^4\theta - 30\cos^2\theta + 3)$	1
42	$\frac{5}{2}(7\cos^2\theta - 1)\sin^2\theta\cos 2\phi$	$\frac{1}{2}\frac{1}{\sqrt{10}}$
44	$\frac{35}{8}\sin^4\theta\cos 4\phi$	$\frac{1}{\sqrt{70}}$
60	$\frac{1}{16}(231\cos^6\theta - 315\cos^4\theta + 105\cos^2\theta - 5)$	1
62	$\frac{105}{32}(33\cos^4\theta - 18\cos^2\theta + 1)\sin^2\theta\cos 2\phi$	$\frac{1}{\sqrt{105}}$
64	$\frac{63}{11}(11\cos^2\theta - 1)\sin^4\theta\cos 4\phi$	$\frac{1}{3\sqrt{14}}$
66	$\frac{231}{32}\sin^6\theta\cos 6\phi$	$\frac{1}{\sqrt{231}}$

wavefunctions and we have no way of determining the potential function at every point in space. The smoothed-out electron distribution can only provide information about parameters $A_n^m\langle r^n\rangle$ with $n \leqslant 6$ for the lanthanide (4f) and actinide (5f) partially filled shells. Possible m values, for a given n, are determined by site symmetry. In the case of the garnets this is D_2, corresponding to three mutually perpendicular two-fold axes, and all positive even values of $m(\leqslant n)$ are allowed. Moreover, in the garnet the lanthanide ion sites have neighbouring oxygen ions at two distances, corresponding to just two values of $\bar{A}_n(R_i)$ for each n.

For $n = 4$ and $n = 6$ this provides a direct test of the superposition model as there are more input data (in the form of the $A_n^m\langle r^n\rangle$ values) than there are parameters. Further tests can be envisaged, such as

(i) checking that the more distant neighbours have smaller interactions,

(ii) comparing the parameters $\bar{A}_n$ obtained for the same neighbouring ions but in crystals of different structure,

(iii) checking to see whether a prediction of first principles calculations, namely $\bar{A}_4 > \bar{A}_6 > 0$, holds in practice.

The coordination factors required for the garnet analysis are given in table (2.1). The sum over i in equation (2.13) has eight terms corresponding to the eight neighbouring O^{2-} ions at the corners of a distorted cube (see figure 1 of the paper in Appendix 2A). The coordination angles and distances are given by Newman and Stedman (Appendix 2A, table 1) or, more completely, by Nekvasil (1979). In order to carry out the calculation it is necessary to evaluate the coordination factors for the four combinations of Θ and Φ corresponding to each value of R_i: (Θ, Φ), $(\Theta, \Phi+\pi)$, $(-\Theta, -\Phi)$, $(-\Theta, -\Phi+\pi)$. It is easy to check that all the formulae in table 2.1 give identical values of the coordination factors, independently of which of the four angular coordinations is concerned. Hence the right-hand side of equation (2.13) is just four times the value determined for a single neighbouring ion at each distance R_1 and R_2.

The aim of the project is to improve upon the analysis given in Appendix 2A, in particular, expected errors in the fitted parameters, and power law exponents should be estimated using the data errors shown in tables III and VI of that paper. Note, however, that although the quoted errors provide an estimate of relative accuracy they do not allow for systematic errors which may occur in fitting the spectroscopic data.

Most students will find it too complicated to introduce angular distortion effects, which are, in any case, not relevant to concentrated paramagnetic crystals. However, it would be worthwhile to examine the accuracy of the x-ray determined ionic positions within the unit cell to see how such errors would reflect on the values of the fitted parameters (see Exercise 2.3). It is still not clear whether the inaccuracies in the fitted parameters are due to a defect in the model or to poor input data.

Note that Nekvasil (1979) uses differently normalised parameters. The relation between the two sets is given in table 2.1.

Exercises (developing Project 2B)

2.2 Is it really necessary or justified to omit the $n=2$ parameters? (See, for example, Newman and Edgar (1976).)

2.3 Write a computer program to evaluate R_i, Θ_i, Φ_i from x-ray data quoted in the usual way (see Euler and Bruce (1965) and Hutchings and Wolf (1964)). Hence determine the sensitivity of the fitting procedure to uncertainties in the x-ray determinations of the ionic positions. Garnets have very complicated structures, but standard crystal models are available.

2.4 What are the specific distortion constraints associated with the rolling ligand model? (See Appendix 2A, p 37.)

2.5 Try to make a realistic assessment of the model in relation to the quality of fits obtained in this project. Do the fits look reasonable? Are consistent values of the parameters $\bar{A}_n$ obtained for different systems? Can a numerical assessment of the quality of fits be made?

PROJECT 2C: THE IONIC MODEL OF CRYSTAL POLARISABILITY

In this project we introduce a simple model of the polarisability of ionic crystals. The model is based on the idea that the electronic excitations which give rise to the experimental polarisability of ionic crystals are localised on the separate ions. We therefore expect the crystal polarisability to be expressible as a sum of ionic polarisabilities. At the same time we must recognise that such polarisabilities may be very different from free ion polarisabilities. A paper by Jaswal and Sharma (reproduced in Appendix 2B), which will be referred to as 'JS' in the following, provides a brief discussion of the theoretical background, input data for the alkali halide crystals and the results of some linear least square fits. It should be consulted in conjunction with the following discussion. Another interesting example, the chalcogenides, has been discussed by Boswarva (1970).

The electronic polarisability α_e of a crystal is determined from the high frequency dielectric constant ε_∞ through the Clausius–Mossotti relation (Boswarva 1970)

$$\frac{\alpha_e}{v} = \frac{3}{4\pi}\left(\frac{\varepsilon_\infty - 1}{\varepsilon_\infty + 2}\right) \tag{2.14}$$

where v is the volume of a unit cell and α_e is the unit cell polarisability. This relation is non-linear because a self-consistency condition must be satisfied, so that the electric field acting on one ion has contributions from the dipole moments induced on the other ions.

We shall follow the notation in JS, using α_i and α_j for the polarisabilities of the positive and negative ions respectively and α_{ij} for the polarisability of the unit cell (which contains just two ions in the alkali halides). The additive nature of the individual ionic polarisabilities corresponds to the equation

$$\alpha_{ij} = \alpha_i + \alpha_j. \tag{2.15}$$

This cannot be written in the form of equation (1.19) with continuous functions $f_\alpha(x)$. Instead, we introduce the logic function $f_\mu^{(ij)}$ which takes the value unity if $\mu = i$ or $\mu = j$ and is zero otherwise. Then equation (1.19) can be expressed in the form

$$\alpha_{ij} = \sum_\mu f_\mu^{(ij)}\, \alpha_\mu. \tag{2.16}$$

In order to obtain the individual ion polarisabilities we may follow the procedure of Tessman *et al* (1953) and minimise the expression

$$S = \sum_{i \neq j} (\alpha_{ij} - \alpha_i - \alpha_j)^2 = \sum_{i \neq j} (\Delta\alpha_{ij})^2$$

where the summation is over all the alkali halide systems for which data are available. In JS there are 20 experimental values of α_{ij} and 9 fitting parameters, α_i, α_j, are used corresponding to the 5 positive ions Li^+, Na^+, K^+, Rb^+ and Cs^+, and the halide ions F^-, Cl^-, Br^- and I^-.

It is instructive to write equation (2.16) in matrix form with $\boldsymbol{y} = (\alpha_{11}, \alpha_{12}, \ldots, \alpha_{21}, \ldots)$ and $\boldsymbol{\theta} = (\alpha_1, \alpha_2, \ldots)$. An abbreviated form of $\mathbf{A}$ for 16 input data and 8 parameters is given below:

$$\mathbf{A} = \begin{bmatrix} 1 & 0 & 0 & 0 & 1 & 0 & 0 & 0 \\ 1 & 0 & 0 & 0 & 0 & 1 & 0 & 0 \\ 1 & 0 & 0 & 0 & 0 & 0 & 1 & 0 \\ 1 & 0 & 0 & 0 & 0 & 0 & 0 & 1 \\ 0 & 1 & 0 & 0 & 1 & 0 & 0 & 0 \\ 0 & 1 & 0 & 0 & 0 & 1 & 0 & 0 \\ 0 & 1 & 0 & 0 & 0 & 0 & 1 & 0 \\ 0 & 1 & 0 & 0 & 0 & 0 & 0 & 1 \\ 0 & 0 & 1 & 0 & 1 & 0 & 0 & 0 \\ 0 & 0 & 1 & 0 & 0 & 1 & 0 & 0 \\ 0 & 0 & 1 & 0 & 0 & 0 & 1 & 0 \\ 0 & 0 & 1 & 0 & 0 & 0 & 0 & 1 \\ 0 & 0 & 0 & 1 & 1 & 0 & 0 & 0 \\ 0 & 0 & 0 & 1 & 0 & 1 & 0 & 0 \\ 0 & 0 & 0 & 1 & 0 & 0 & 1 & 0 \\ 0 & 0 & 0 & 1 & 0 & 0 & 0 & 1 \end{bmatrix} \tag{2.17}$$

Using this matrix it is easy to show that the $M \times M$ matrix $\mathbf{A}^{\mathrm{T}}\mathbf{W}\mathbf{A}$ is singular for *any* diagonal matrix $\mathbf{W}$. The reason for this is that we can transform the parameters, $\alpha_i \rightarrow \alpha_i + c$, $\alpha_j \rightarrow \alpha_j - c$, without altering the fit. It is thus necessary to assume a value for one of the ionic polarisabilities. This is incorporated into the fit by providing the matrix $\mathbf{A}$ with an additional row: e.g. $[1 \quad 0 \quad 0 \quad 0 \quad 0 \quad 0 \quad 0 \quad 0]$ if the value of α_1 is assumed. In fitting the alkali halide data it is usual to take the smallest positive ion to have the calculated free ion value $\alpha(Li^+) = 0.029$ Å. Note that it is quicker to write a procedure to generate the matrix $\mathbf{A}$ than it is to enter the whole matrix, element by element.

Pirenne and Kartheuser (1964) have suggested that it would be better to minimise the expression

$$\begin{aligned} S' &= \frac{1}{N} \sum_{i \neq j} \left(\alpha_{ij} - \sum_{\mu} f_{\mu}^{(ij)} \alpha_{\mu} \right)^2 \alpha_{ij}^{-2} \\ &= \frac{1}{N} \sum_{i \neq j} \left(1 - \sum_{\mu} \frac{f_{\mu}^{(ij)} \alpha_{\mu}}{\alpha_{ij}} \right)^2 \end{aligned} \tag{2.18}$$

so as to give the smallest relative polarisability deviations, $\Delta\alpha_{ij}$. If we put $\boldsymbol{y} = (1 \quad 1 \quad 1 \ldots 1)^{\mathrm{T}}$ and write $f_{\mu}^{(ij)}/\alpha_{ij}$ instead of $f_{\mu}^{(ij)}$, it is easy to modify the fitting procedure described above to minimise S' rather than S.

It should be remarked that the total electric polarisability of ionic crystals depends on the sum of the effect of the electronic excitations and the displacement of the ions from their equilibrium lattice sites. The additive model used in this project represents only the contribution of the electronic excitations. It is separated experimentally from the ionic displacement contribution by applying electric fields of such high frequency that the slow-moving ions are unable to respond.

The model described above can be generalised in various ways to allow for some correlation between the polarisabilities of the two types of ion making up the crystal. Pirenne and Kartheuser (1964) have suggested the introduction of an additional parameter λ; we can then write

$$\alpha_{ij} = \alpha_i + \alpha_j + \lambda\alpha_i\alpha_j. \tag{2.19}$$

An alternative suggestion (P M Hui, private communication) is to write

$$\alpha_{ij} = \alpha_i + \alpha_j + \lambda\sqrt{\alpha_i\alpha_j}. \tag{2.20}$$

This has the advantage that the additional parameter is dimensionless. The main question is, of course, whether either or both or these generalisations give a better fit to the data than simple additivity.

In developing the model, important questions also arise in relation to finding specific arguments to justify the form of equations (2.19) and (2.20). Such justifications are not provided in the literature, but may be related to the 'shell model' expression for the electronic polarisability (Dick and Overhauser 1958). An important aspect of this relates to the possibility of relating the ionic polarisability model to other types of experimental data. The connection with the shell model is particularly interesting in this regard as this model has provided a successful parametrisation of the lattice vibration frequencies for a wide range of different crystals (see, for example, Peckham (1967) and Sangster *et al* (1970) for a discussion of MgO).

It is not possible to carry out a *linear* least squares fit to equations (2.19) and (2.20). We may, however, iterate towards a solution using the formula

$$\alpha_{ij} = \alpha_i^{(n)} + \alpha_j^{(n)} + \lambda^{(n)}\alpha_i^{(n-1)}\alpha_j^{(n-1)}.$$

The initial iteration is the linear least squares fit

$$\alpha_{ij} = \alpha_i^{(1)} + \alpha_j^{(1)}$$

instead of equation (2.19). The superscripts $(n-1)$ and (n) indicate the order of the iteration. The nth iteration is now a linear least squares fit to the parameters with superscript (n), while holding the parameters with superscript $(n-1)$ constant. Iteration proceeds until $\alpha_i^{(n)} = \alpha_i^{(n-1)}$. In carrying out any non-linear least squares fit it is necessary to be wary of false minima. A common precaution is to try various sets of initial values of the parameters.

A general linear least squares fitting program should be easy to adapt to this project. It is only necessary to insert a routine which will generate the matrix **A**. Iteration can be carried out interactively or by inserting an additional loop into the program. Interactive iteration has the advantage of allowing interference with the starting values of each iteration, such as might be necessary to damp out fluctuations in successive sets of parameter values.

It should be remarked that the model described in this example is not universally accepted. Pantelides (1975) has proposed a model based on the extreme opposite hypothesis: that the crystal polarisability depends entirely on virtual excitations from the negative ion to the positive ion. His model also leads to a simple formula which appears to agree well with the experimental data.

Exercises (developing Project 2C)

2.6 Show that $\mathbf{A}^T\mathbf{W}\mathbf{A}$ is singular for the matrix $\mathbf{A}$ given by equation (2.17).

2.7 Find the parameter transformations which do not alter the fit to the non-linear equations (2.19) and (2.20). The existence of such transformations shows that an extra row in $\mathbf{A}$ is also necessary when carrying out non-linear fits.

2.8 Compare and contrast the predictive power of the Pantelides (1975) model with the model described in these project notes.

PROJECT 2D: LIQUID-DROP MODEL OF NUCLEI

In this model the atomic nucleus is treated simply as an unstructured collection of nucleons, all nuclei having the same nucleon density (i.e. consisting of droplets of the same liquid). There are two types of interaction which affect the binding energy of such a system, the short-range (attractive) nuclear force acting between all neighbouring nucleons and the long-range (repulsive) Coulomb force acting between protons.

The nuclear force affects the total energy of the nucleus in two ways. One contribution is proportional to the volume of the nucleus and is based on the assumption that every nucleon has (on average) an equal number of interacting neighbours. As the density is fixed, this contribution is proportional to the number of nucleons (A) in the nucleus and may be written

$$\Delta E_{\text{vol}} = -a_1 A \tag{2.21}$$

where the constant $a_1 > 0$. This expression for the binding energy must be modified to allow for the fact that nucleons on the surface have fewer neighbours. If A is proportional to the volume, then $A^{2/3}$ will be proportional to the surface area. Hence the surface correction may be written

$$\Delta E_{\text{surf}} = a_2 A^{2/3} \tag{2.22}$$

where $a_2 > 0$ parametrises the reduction in the volume-dependent contribution.

The Coulomb repulsion acts between pairs of the Z protons, each pair giving a (positive) potential energy contribution proportional to $1/R$ where R is the distance between them. The mean distance will be proportional to a characteristic distance in the nucleus, say its radius, and hence proportional to $A^{1/3}$. There are $Z(Z-1)/2$ interactions between Z protons but this is often replaced by the expression $Z^2/2$, which is appropriate for the self-interaction of a continuous charge distribution. We may therefore write the Coulomb repulsion term as

$$\Delta E_C = a_3 Z^2 A^{-1/3} \quad \text{or} \quad a_3 Z(Z-1) A^{-1/3} \tag{2.23}$$

where $a_3 > 0$ corresponds to a repulsion effect.

Consider now a nucleus with a fixed number of nucleons (A) but variable proportions of protons (Z) and neutrons ($A-Z$). The Coulomb correction would lead us to predict that the most stable nucleus is made up only of neutrons, but this is contrary to the experimental evidence which shows nuclei with a similar number of protons and neutrons to be the most stable. A term which gives this effect is

$$\Delta E_{sym} = a_4 \frac{(Z - 0.5A)^2}{A} \tag{2.24}$$

with $a_4 > 0$, corresponding to a repulsive potential proportional to the square of the deviation of the system from the symmetrical configuration with $Z = A - Z$. This provides a simple way of allowing for the effect of the Pauli exclusion principle within the framework of the liquid-drop model.

It is, of course, quite possible to generate simple models in which the constants a_1, a_2, a_3 and a_4 can be related to other nuclear properties and hence calculated. Such calculations are, however, not found to be satisfactory. In this situation the best procedure is simply to fit the four parameters a_1, a_2, a_3 and a_4, defined by equations (2.21) to (2.24) above, to experimental values of the binding energy (i.e. the empirical nuclear mass defects).

In recent years the liquid-drop model of nuclei has been developed further, both by introducing additional terms and corrections (such as for the electron masses) and by using other data in the fitting procedure (such as stability with respect to β-decay). The liquid-drop model has been found to be particularly useful for the description of nuclear fission.

Notes and Comments

There are many other minor corrections that could be taken into account in fitting the mass defect formula

$$\Delta E = \Delta E_{vol} + \Delta E_{surf} + \Delta E_C + \Delta E_{sym}$$

to the empirical data for nuclear masses. We shall not go into these details here. The interested reader is referred to Green (1955). Lists of nuclear masses may be found in standard reference works, such as the *CRC Handbook of Chemistry and Physics* (1975) and these can be used to obtain the values of a_n. It will be apparent that the four terms in the expression for ΔE are not orthogonal, so some thought should be given to the best procedure and the best data to use. Green (1955, p 287) tabulates several sets of values of the a_n coefficients which have been determined by different workers. Overall, the values to be expected are

$$a_1 = 16 \pm 1 \qquad a_2 = 17 \pm 3 \qquad a_3 = 0.7 \pm 0.1 \qquad a_4 = 90 \pm 10.$$

The non-orthogonality of the fitting functions does, of course, mean that there is considerable correlation between the values obtained for the different parameters. It would be interesting to analyse the values tabulated by Green to see whether orthogonalisation would give better agreement between some of the different sets of fitted parameters.

References

Abragam A and Bleaney B 1970 *Electronic Paramagnetic Resonance of Transition Ions* (Oxford: Clarendon) p 567, 569
Boswarva I M 1970 *Phys. Rev.* B **1** 1698
Dick Jr B G and Overhauser A W 1958 *Phys. Rev.* **112** 90
Euler F and Bruce J A 1965 *Acta Crystallogr.* **19** 971
Fish G E, North M H and Stapleton H J 1980 *J. Chem. Phys.* **73** 4807
Green A E S 1955 *Nuclear Physics* (New York: McGraw-Hill)
Handbook of Chemistry and Physics 1975 56th edn Table of the Isotopes B252–336 (Cleveland, Ohio: Chemical Rubber)
Hutchings M T and Wolf W P 1964 *J. Chem. Phys.* **41** 617
Jackson J D 1962 *Classical Electrodynamics* (New York: Wiley) p 69
Jones A 1970 *Comput. J.* **13** 301
Kurkin I N and Tsvetkov E A 1970 *Sov. Phys.–Solid State* **11** 3027
Lawson C L and Hanson R J 1974 *Solving Least Square Problems* (Englewood Cliffs, NJ: Prentice-Hall)
Nekvasil V 1979 *Czech. J. Phys.* B **29** 785
Newman D J 1971 *Adv. Phys.* **20** 197
—— 1978 *Aust. J. Phys.* **31** 489
Newman D J and Edgar A 1976 *J. Phys. C: Solid State Phys.* **9** 103
Orbach R 1961a *Proc. Phys. Soc.* **77** 821
—— 1961b *Proc. R. Soc.* A **264** 458
—— 1962 Spin-Lattice Relaxation in Solids in *Fluctuation, Relaxation and Resonance in Paramagnetic Systems* ed. D ter Haar p 219–29 (Edinburgh: Oliver and Boyd)

Pantelides S T 1975 *Phys. Rev. Lett.* **35** 250
Peckham G 1967 *Proc. Phys. Soc.* **90** 657
Pirenne J and Kartheuser E 1964 *Physica* **30** 2005
Sangster M J L, Peckham G and Saunderson D H 1970 *J. Phys. C: Solid State Phys.* **3** 1026
Tessman J R, Kahn A H and Shockley N 1953 *Phys. Rev.* **92** 890

Appendix 2A: Interpretation of Crystal-Field Parameters in the Rare-Earth-Substituted Garnets†

D. J. NEWMAN AND G. E. STEDMAN

Department of Physics, Queen Mary College, Mile End Road, London, E.1., United Kingdom

(Received 17 March 1969)

Crystal-field parameters of Yb^{3+}, Eu^{3+}, Tm^{3+}, Sm^{3+}, Dy^{3+}, and Er^{3+} ions in various garnet lattices have been analyzed using a development of the superposition model proposed by Bradbury and Newman. This effects, in principle, the separation of the crystal-field contributions of the different coordinated O^{2-} ions and determines power laws and the angular distortion near substituted ions. A realistic separation of the contributions to the $n=4$ parameters is obtained even when no allowance is made for local distortion. However, owing to their greater angular sensitivity, the contributions to the $n=6$ parameters can only be separated if local angular distortions of up to 9° are allowed for. Results derived for single-ligand parameters and power laws are in good accord with theoretical results obtained for other rare-earth systems.

I. INTRODUCTION

It has been suggested previously[1] that x-ray data could be used in conjunction with the "superposition approximation" to determine geometrical constraints on the crystal-field-parameter ratios A_n^m/A_n^0, and that such a technique may be used to reduce the number of free parameters which are required to fit the spectroscopic data in the case of low symmetry sites. In those cases where a complete set of parameters for a low symmetry site with coordinated ions at different distances already exists, it should also be possible to extract information about the effective single-ligand crystal-field parameters for different metal–ligand distances. The situation is complicated, however, by the distortion that may occur in the immediate neighborhood of a substituted ion. Yet even this may be turned to advantage if sufficient experimental data is available to enable us to *deduce* constraints on the form of these distortions.

The purpose of this paper is to initiate a program of work on these lines for rare-earth ions substituted into the garnets. The main difficulties encountered are due to the incomplete and rather uncertain experimental crystal-field parameters at present available for the garnets and the problems involved in constructing an adequate model of the local distortion with sufficiently few undetermined parameters. However, our results are sufficiently encouraging to justify further effort in both of these directions.

In Sec. II we describe the crystalline structure of the garnets in the region of the rare-earth ions. In Secs. III–V the application of the superposition approximation to the garnet system and the methods of fitting to the experimental data are described. Results are discussed in Sec. VI and general conclusions are drawn in Sec. VII.

II. STRUCTURE OF THE GARNETS

The x-ray data of Euler and Bruce[2] show that Y^{3+} and Lu^{3+} ions in the garnets are found at sites of D_2 symmetry, with eight coordinated[3] O^{2-} ions (see Fig. 1). The latter are situated at the corners of a distorted cube centered on the Y^{3+} or Lu^{3+} ion and at two distinct distances, which we denote R_1, R_2. Values of these constants and the angular positions of the O^{2-} ions in the different host crystals are given in Table I.

Following Hutchings and Wolf,[4] where more details and tables of x-ray data are given, we employ a local coordinate system at the cation site $(\frac{1}{8}, 0, \frac{1}{4})$ with axes coincident with the twofold symmetry axes [1, 0, 0], [0, $\bar{1}$, $\bar{1}$], [0, $\bar{1}$, 1]. Koningstein[5] employs axes which are inclined to ours by an angle of 45°; this merely changes the signs of the crystal-field parameters $A_4^2\langle r^4\rangle$, $A_6^2\langle r^6\rangle$, and $A_6^6\langle r^6\rangle$.

III. APPLICATION OF THE SUPERPOSITION APPROXIMATION TO THE GARNET STRUCTURE

We refer the reader to previous work for a full (mathematical) description of the *superposition approximation*.[1,6,7] In this approximation the total crystal field is assumed to be the sum of independent contributions from each ion in the neighborhood of the substituted rare-earth ion. Because of the known inadequacy of the electrostatic-model, contributions of the coordinated O^{2-} ions are parametrized so as to allow for the effects of overlap, covalency, and shielding without performing an elaborate *ab initio* calculation. The uncoordinated ions are more difficult to handle, as we do not wish to introduce extra parameters, and, although their major contribution to the crystal field is certainly electrostatic, this will be reduced by the shielding effect of the polarized $5s^2p^6$ shell. No precise calculations (including

[1] M. I. Bradbury and D. J. Newman, Chem. Phys. Letters **1**, 44 (1967).

[2] F. Euler and J. A. Bruce, Acta Cryst. **19**, 971 (1965).

[3] The word "coordinated" is used to indicate that there are no intervening ions. We avoid the term "nearest neighbor" which is sometimes used in a stricter sense.

[4] M. T. Hutchings and W. P. Wolf, J. Chem. Phys. **41**, 617 (1964).

[5] J. A. Koningstein in *Optical Properties of Ions in Crystals*, H. M. Crosswhite and H. W. Moos, Eds. (Interscience Publishers, Inc., New York, 1967), pp. 108–109.

[6] G. E. Stedman, J. Chem. Phys. **50**, 1461 (1969); also "Interpretation of the Crystal Field for V^{2+}, Cr^{3+}, and Mn^{4+} in Corundum," J. Chem. Phys. (to be published).

[7] M. M. Curtis, D. J. Newman, and G. E. Stedman, J. Chem. Phys. **50**, 1077 (1969).

†Reprinted from *J. Chem. Phys.* 1969 **51**(7) 3013–23.

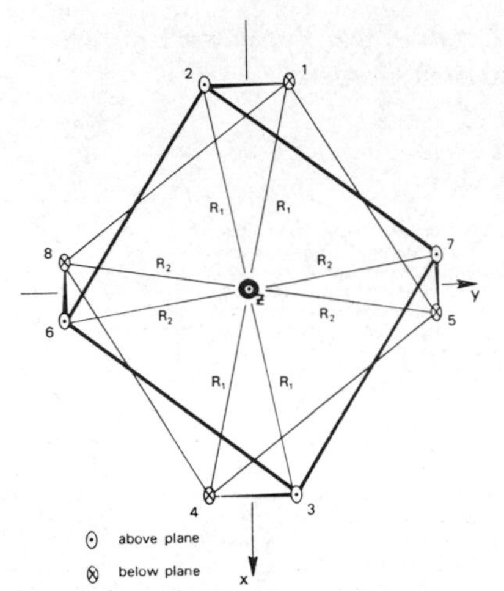

FIG. 1. The complex $(YO_8)^{13-}$ in YGG. Oxygen–oxygen distances kept constant in the rolling-ligand model are as follows: $d_{12}=d_{34}=2.81$ Å; $d_{15}=d_{26}=d_{37}=d_{48}=2.66$ Å; $d_{18}=d_{27}=d_{36}=d_{45}=2.78$ Å; $d_{57}=d_{68}=2.96$ Å. (*Note added in proof:* the figure should be reflected through the zx plane.)

overlap and covalency contributions) exist on the extent of this effect. A rough indication of the relative importance of the uncoordinated ions is obtained by comparing the (unshielded) electrostatic point-charge contributions, as calculated by Hutchings and Wolf,[4] with the experimental $n=4$ and $n=6$ parameters (see, for example, Tables II and III). The $n=2$ parameters will not be considered in our analyses as, in this case, the distant-ion electrostatic contributions are probably dominant.

The crystal field due to each coordinated ion (or ligand) is assumed to be axially symmetric, and is thus parametrized by three coefficients $\bar{A}_n(R)$ $(n=2, 4, 6)$ of the tesseral harmonics $Z_0{}^n$, where R is the distance between the ligand and rare-earth ion. It can then be shown (see Refs. 1 and 7) that the total crystal field has static parameters

$$A_n{}^m\langle r^n\rangle=\sum_T(S_m{}^n/K_0{}^n)Z_m{}^n(\Theta_T,\Phi_T)\bar{A}_n(R_T),$$

where the suffix T refers to the different ligands, at positions (R_T,Θ_T,Φ_T) and the numerical factors $S_m{}^n$ and $K_0{}^n$ have been defined previously (Ref. 7, Table VIII). In cases where the same ligand occurs at various distances it is also useful to define an effective *power law* $t_n=-R(\partial\bar{A}_n/\partial R)(\bar{A}_n)^{-1}$ which, it is hoped, will be fairly independent of R in the restricted range considered.

As the coordinated O^{2-} ions are at two distances in the garnets, six axial parameters $\bar{A}_n(R_i)$ $(i=1, 2)$ are required. These will be referred to as the *intrinsic parameters* of the complex. The angular coordination of the O^{2-} ions relates these to the nine experimentally determined parameters $A_n{}^m\langle r^n\rangle$ $(m\leqq n=2, 4, 6;$ $m=0, 2, 4, 6)$ which describe the crystal field of a rare-earth ion in $\mathbf{D}_2$ symmetry. Restricting ourselves to $n=4, 6$, there are four intrinsic parameters to relate to the seven experimental ones if we assume that the angular positions of the ligands may be derived from the known structure of the host crystal. However, it is known that small but significant changes of angular coordination can take place round a substituted ion. We may therefore attempt to use the excess of experimental information to determine the extent of the local distortion in the garnets by devising simple models which must have at most three parameters.

The most direct method would be to regard the four angles which describe the angular positions of ions 1 and 5 as free parameters (the positions of the remaining ions in the complex being determined by the site symmetry). However, this requires one more experimental parameter than is available. Investigation of this fitting space for several of the garnet systems suggest that the major local distortion is a change in the angular position of the more distant ions. We have therefore tried several fittings using a model, to be referred to as the *angular-distortion model,* in which only the position (θ_2, φ_2) of ion 5 is varied. It should be remarked that, although metal–ligand distance variations do not appear explicitly in this model, the appropriate degrees of freedom are in fact absorbed by the variation of intrinsic parameters.

The main disadvantage of the angular-distortion model is that it gives no indication of the associated radial distortion which occurs and thus leaves some uncertainty in the derived power laws, although their relative magnitude should be realistic. We therefore introduce another model, to be called the *rolling-ligand model,* in which the four smallest oxygen–oxygen distances (specified by the figure) are assumed to remain

TABLE I. Positions of the O^{2-} ligands in undistorted host crystals. $(R_1\theta_1\varphi_1)$ refers to ion 1 and $(R_2\theta_2\varphi_2)$ to ion 5 in the figure.

	R_1 (Å)	R_2 (Å)	θ_2	φ_2	θ_1	φ_1
YAG	2.3030	2.4323	125.94°	+81.24°	123.86°	−192.52°
YGG	2.3383	2.4277	126.69°	+80.90°	125.33°	−191.59°
LuGG	2.3025	2.3927	127.20°	+80.59°	125.84°	−191.30°

TABLE II. Relative expected accuracies used as weights in the fitting procedures and corrections made for distant-ion electrostatic contributions in the compensated fits (cm^{-1}).

	Relative expected accuracies	Distant-ion electrostatic contributions			
		Er^{3+}:YGG	Er^{3+}:YAG	Dy^{3+}:YGG	Dy^{3+}:YAG
$A_4^0\langle r^4\rangle$	8.3	6.1	6.6	7.2	7.8
$A_4^2\langle r^4\rangle$	48	65	75	77	88
$A_4^4\langle r^4\rangle$	42	−93	−106	−109	−124
$A_6^0\langle r^6\rangle$	5.4	0.5	0.6	0.6	0.7
$A_6^2\langle r^6\rangle$	60	−7	−9	−10	−11
$A_6^4\langle r^6\rangle$	45	12	14	15	18
$A_6^6\langle r^6\rangle$	42	−14	−16	−18	−21

constant at their x-ray values in the pure salt. This is based on the expectation that the O^{2-}–O^{2-} bonds will be less compressible than the *metal*–O^{2-} bonds. The six parameters which describe an arbitrary distortion which retains $\mathbf{D}_2$ symmetry are thus reduced to two, while retaining the useful feature that changes in the metal–ligand distances R_1 and R_2 are predicted.

All analyses of the experimental data have been carried out using the least-squares technique developed by Powell[8] which is found to be much faster than the simple methods. The weight of each experimental parameter in the fitting procedures was determined by dividing by its expected accuracy as derived from the quoted errors in the Yb^{3+}:YGG data (see Table II). These errors are, in fact, mainly due to the different normalization of the parameters so it is not unrealistic to use the same weighting in the fits to all sets of experimental parameters.

It should be noticed that, in the undistorted fits, a linear relationship exists between the data and fitted parameters, but the data is nonlinear in the distortion parameters. In both models which allow for distortion there will thus exist the possibility of hidden constraints on the possible range of the crystal-field parameters that may be generated.

IV. CRITERIA FOR A GOOD FIT TO THE EXPERIMENTAL DATA

In those analyses where it is assumed that the host lattice is undistorted, the $n=4$ and $n=6$ parameters

TABLE III. An analysis of Er^{3+}:YGG parameters ($A_n^m\langle r^n\rangle$ and $\bar{A}_n$ are given in cm^{-1}).

	Experiment	No distortion	Angular distortion	Angular distortion (compensated)
$A_4^0\langle r^4\rangle$	−238±0.5	−246	−238	−241
$A_4^2\langle r^4\rangle$	255±2	254	255	252
$A_4^4\langle r^4\rangle$	920±2	860	916	902
$\bar{A}_4(R_1)$		96.6	95.6	93.6
$\bar{A}_4(R_2)$		58.9	55.4	64.3
t_4		13.2	14.5	10.0
$A_6^0\langle r^6\rangle$	33.0±0.1	36	33	36
$A_6^2\langle r^6\rangle$	−58±2	−24	−58	−67
$A_6^4\langle r^6\rangle$	645±1	636	642	632
$A_6^6\langle r^6\rangle$	−70±2	−31	−70	−85
$\bar{A}_6(R_1)$		22.8	23.1	23.6
$\bar{A}_6(R_2)$		22.3	18.1	16.8
t_6		0.5	6.6	9.0
$\Delta\theta_2$			−0.2°	−1.1°
$\Delta\varphi_2$			9.1°	8.9°

[8] M. J. D. Powell, Computer J. **7**, 303 (1965); Harwell Subroutine Library program VA02A.

TABLE IV. Analysis of ErGG parameters ($A_n^m\langle r^n\rangle$ and $\bar{A}_n$ are given in cm^{-1}).

	Experiment	No distortion	Angular distortion	Angular distortion (compensated)	Rolling-ligand model
$A_4^0\langle r^4\rangle$	−221	−240	−226	−226	−225
$A_4^2\langle r^4\rangle$	246	243	239	238	249
$A_4^4\langle r^4\rangle$	978	839	948	952	954
$\bar{A}_4(R_1)$		94.0	88.8	85.8	95.6
$\bar{A}_4(R_2)$		57.6	58.2	68.6	48.6
t_4		13.0	11.3	6.0	−14.6
$A_6^0\langle r^6\rangle$	31.5	36	38	40	26
$A_6^2\langle r^6\rangle$	−71	−41	−77	−94	−42
$A_6^4\langle r^6\rangle$	657	640	625	614	673
$A_6^6\langle r^6\rangle$	−55	−26	−69	−82	−12
$\bar{A}_6(R_1)$		23.7	25.7	26.3	17.5
$\bar{A}_6(R_2)$		21.7	16.0	14.8	22.5
t_6		2.4	12.6	15.3	5.5
$\Delta\theta_2$			−2.3°	−3.4°	
$\Delta\varphi_2$			9.1°	9.1°	
R_1(Å)	2.3383				2.4520
R_2 (Å)	2.4277				2.3411

are treated separately, so that we may evaluate the effectiveness of the procedure independently for $n=4$ and $n=6$. In those analyses which allow for distortions, however, the experimental parameters are used collectively to determine the optimum distortion and we can evaluate the results only of the over-all procedure.

Several criteria may be employed to test the effectiveness of the fittings and hence the applicability of the superposition approximation to the garnets:

(i) Only small residual errors should occur between the fitted and experimental parameters.

(ii) The values of the derived $n=4$, 6 intrinsic parameters should always be positive. This is a definite prediction of *ab initio* theory and is known to be true on the basis of experimental data for rare-earth ions substituted into $LaCl_3$.[7] We also expect the intrinsic parameters to have a reasonable relationship to those found experimentally and theoretically in other systems. In particular, the various cases examined suggest that the ratio $\bar{A}_4/\bar{A}_6$ should lie between the approximate limits 2 and 4.

(iii) The values of t_n should be positive, corresponding to a decrease of the parameters with increasing distance. In this case no relevant experimental data exists, but the theory for Pr^{3+}–Cl^- [5] gives $t_6\sim11$, $t_4\sim10$, while that for Pr^{3+}–F^- [9] gives $t_6=5.6$, $t_4=5.7$. This makes prediction a bit chancy, but at least we expect consistent results with $t_4\sim t_6$ which are fairly independent of the garnet host crystal and substituted rare-earth ion.

(iv) In those analyses which allow for local distortion this should be sufficiently small to be realistic.

V. RESULTS

We have carried out analyses of all known sets of spectroscopic data for rare-earth ions in paramagnetic garnets where the number of quoted crystal-field parameters is sufficient to give definite results. These are presented individually below, but first we make some general observations.

The accuracy of the parameters varies considerably and, where several sets of data were available, we have analyzed only the most accurate. Explicit statements of errors are available only for Er^{3+}:YGG, Er^{3+}:LuGG, and Yb^{3+}:YGG parameters. Many of the fits to experimental parameters discussed below are inaccurate, and there is a consequent uncertainty in the results obtained because these then depend on the choice of weighting used in the least-squares procedure. In all cases other than Sm^{3+} the accuracy of the fit which allows for angular distortion of the complex is within a reasonable estimate of the accuracy of the experimental results (given that x-ray data is available only for YGG, YAG, and LuGG).

Each table of results is arranged so that the first column contains the experimental parameters used in

[9] D. J. Newman and M. M. Curtis, "Crystal Field in Rare-Earth Fluorides. I. Molecular Orbital Calculation of PrF_3 Parameters" J. Phys. Chem. Solids (to be published).

the analysis and the second column is a fit to these parameters without allowing for local distortions in the complex. A third column then gives a fit which is based on a one- or two-parameter angular distortion of the rare-earth complex according to whether φ_2 alone or both θ_2 and φ_2 are allowed to vary. The distinction is indicated in the tables by whether or not a numerical value is recorded for $\Delta\theta_2$. In those systems where only six of the seven crystal-field parameters are available, a one-parameter angular distortion is used. Further columns are used to give "compensated" angular-distortion fits in which the electrostatic contributions due to uncoordinated ions are allowed for, and "rolling-ligand" model fits based on the constrained distortion of the complex described in Sec. III.

A. Er^{3+}:YGG and ErGG

Relatively accurate experimental data were available for these systems,[10,11] the main differences being in the parameters $A_6^2\langle r^6\rangle$ and $A_6^6\langle r^6\rangle$, where they exceed 20%. The x-ray data used was for YGG in both cases, as data for ErGG is not available. When angular distortions are allowed for, the fit to experimental data supplied by Crosswhite (Table III) is very much better

TABLE V. Analysis of ErAG parameters ($A_n^m\langle r^n\rangle$ and $\bar{A}_n$ are given in cm^{-1}).

	Experiment	No distortion	Angular distortion	Angular distortion (compensated)
$A_4^0\langle r^4\rangle$	−233	−253	−242	−243
$A_4^2\langle r^4\rangle$	260	256	250	245
$A_4^4\langle r^4\rangle$	1080	943	1030	1023
$\bar{A}_4(R_1)$		107.2	103.0	100.0
$\bar{A}_4(R_2)$		60.0	59.5	70.7
t_4		10.6	10.1	6.4
$A_6^0\langle r^6\rangle$	36.0	44	44	45
$A_6^2\langle r^6\rangle$	−108	−59	−89	−104
$A_6^4\langle r^6\rangle$	666	634	627	615
$A_6^6\langle r^6\rangle$	−90	−47	−93	−109
$A_6(R_1)$		25.2	26.2	26.5
$\bar{A}_6(R_2)$		22.0	17.5	16.3
t_6		2.5	7.4	8.8
$\Delta\theta_2$			−1.3°	−2.3°
$\Delta\varphi_2$			8.8°	8.7°

[10] H. M. Crosswhite and H. W. Moos, in *Optical Properties of Ions in Crystals*, H. M. Crosswhite and H. W. Moos, Eds. (Interscience Publishers, Inc., New York, 1967), p. 28. Corrections and error estimates have also been supplied by H. M. Crosswhite (private communication).

[11] P. Grunberg, S. Hüfner, E. Orlich, and J. Schmitt, J. Appl. Phys. **40**, 1501 (1969); E. Orlich and S. Hüfner, *ibid.* **40**, 1503 (1969); also E. Orlich (private communication).

TABLE VI. Analysis of Er^{3+}: LuGG parameters ($A_n^m\langle r^n\rangle$ and $\bar{A}_n$ are given in cm^{-1}).

	Experiment	No distortion	Angular distortion
$A_4^0\langle r^4\rangle$	−245±0.5	−256	−245
$A_4^2\langle r^4\rangle$	175±2	175	175
$A_4^4\langle r^4\rangle$	948±2	864	946
$\bar{A}_4(R_1)$		93.2	90.7
$\bar{A}_4(R_2)$		65.9	63.1
t_4		9.0	9.5
$A_6^0\langle r^6\rangle$	33.8±0.1	34.6	34.1
$A_6^2\langle r^6\rangle$	−83±2	−54	−86
$A_6^4\langle r^6\rangle$	664±2	663	662
$A_6^6\langle r^6\rangle$	−48±2	−12	−52
$\bar{A}_6(R_1)$		24.9	25.8
$\bar{A}_6(R_2)$		21.9	16.9
t_6		3.4	11.1
$\Delta\theta_2$			−0.9°
$\Delta\phi_2$			7.9°

than that obtained to Orlich's data (Table IV). In fact, the fit to Crosswhite's parameters bears a very reasonable relationship to his quoted errors. We shall take these errors literally and assign the discrepancies in our fit to Orlich's parameters mainly to the inadequate x-ray data. This will be seen to be consistent with our later results.

The predicted distortions and mean intrinsic parameters {i.e., $\frac{1}{2}[A_n(R_1)+A_n(R_2)]$} are similar for the two sets of data, but the derived power laws are very different. This is found to be a general feature of our results; the power laws obtained vary considerably and unsystematically from case to case, while the predicted distortion favors $\Delta\theta_2 \sim 0°$ and $\varphi_2 \sim 90°$, corresponding to the unexpectedly large value of $\Delta\varphi_2 \sim 9°$.

Angular distortion fits have also been carried out allowing for distant-ion electrostatic contributions based on the lattice sums of Hutchings and Wolf.[4] Results for these "compensated" fits are also given in Tables III and IV. In the case of Crosswhite's data the previously good fit is spoilt and no over-all improvement is obtained in the case of Orlich's parameters. This implies that either the lattice sum gives an inadequate representation of the electrostatic field or that this field is effectively quenched by screening effects in the rare-earth ion and its coordinated oxygens.

An attempt has also been made to use the "rolling-ligand" model to analyze Orlich's data (Table IV). This failed completely, as the relative magnitudes of R_1 and R_2 are reversed, and a negative value of t_4 is predicted.

TABLE VII. Analysis of Dy^{3+}:YGG parameters ($A_n^m\langle r^n\rangle$ and $\bar{A}_n$ are given in cm^{-1}).

	Experiment	No distortion	Angular distortion	Angular distortion (compensated)
$A_4^0\langle r^4\rangle$	−282 (−235)	−287	−282	−284
$A_4^2\langle r^4\rangle$	181 (215)	180	181	179
$A_4^4\langle r^4\rangle$	1040 (945)	1004	1039	1026
$\bar{A}_4(R_1)$		105.8	106.9	104.8
$\bar{A}_4(R_2)$		75.1	69.6	78.9
t_4		9.1	11.4	7.6
$A_6^0\langle r^6\langle$	36 (67)	41	36	38
$A_6^2\langle r^6\rangle$	−108 (−10)	−71	−105	−119
$A_6^4\langle r^6\rangle$	725 (902)	709	726	715
$A_6^6\langle r^6\rangle$	−57 (−200)	−19	−53	−74
$\bar{A}_6(R_1)$		27.7	27.2	27.7
$\bar{A}_6(R_2)$		22.9	20.1	17.9
t_6		5.1	8.1	11.5
$\Delta\theta_2$			0.9°	0.0°
$\Delta\varphi_2$			5.7°	9.1°

B. ErAG

The experimental data for this system (Table V) was supplied by Orlich.[11] The angular-distortion fit is of similar quality to that obtained with Orlich's ErGG data. Some uncertainty therefore exists in the derived values of t_4 and t_6. Again, the introduction of distant electrostatic contributions in the compensated fit does not improve agreement with the experimental data.

C. Er^{3+}:LuGG

The experimental data in this case was supplied by Crosswhite.[10] The results quoted in Table VI show that here, just as in the case of Crosswhite's Er^{3+}:YGG data, the two-parameter angular-distortion fit bears a reasonable relationship to the stated errors.

D. Dy^{3+}:YGG

The results given in Table VII are derived from experimental parameters obtained by Grunberg *et al.* and supplied by Orlich.[11] The earlier, approximate parameters given in brackets are due to Veyssie.[12]

The two-parameter angular-distortion fit is very close to the experimental parameters considering the expected accuracy of Orlich's data, but the predicted distortion is rather less than that found in the previous cases. When the parameters are compensated for the electrostatic contributions of the uncoordinated ions, the fit becomes rather worse, but still within a reasonable estimate of the accuracy of the data. A distortion which is more consistent with previous findings is also obtained.

E. Dy^{3+}:YAG

The analysis for this system is based on experimental parameters obtained by Grunberg *et al.* and supplied by Orlich,[11] and is given in Table VIII. Although the fits to $n=4$ parameters are satisfactory, the angular-distortion fits do not give very satisfactory results for $n=6$ as compared with the previous case. The rolling-ligand model gives an extreme, and therefore unphysical, distortion.

F. Sm^{3+}:YGG and Sm^{3+}:YAG

Both sets of data (Tables IX and X) were obtained by Grunberg *et al.* and supplied by Orlich.[11] They were obtained from ^6H term splittings alone, as it was found that a distinct set of parameters were required to represent the ^6F term splittings. This is a direct indication that correlation crystal-field effects[13] are important in Sm^{3+}, and warns us that there may be difficulties in applying our model in these cases.

These reservations are borne out in practice, the fits being well outside the expected errors. In particular it will be noticed that in no case are we able to reproduce the negative sign of $A_6^6\langle r^6\rangle$. Also the only reasonable fit to the $n=4$ parameters in SmGG leads to a negative value of t_4. The two-parameter fits cannot therefore be used as a source of useful information. Only in the one-parameter angular-distortion fits are convincing results obtained, but these must clearly be treated with caution

[12] M. Veyssie, Ph.D. thesis, Université de Grenoble, 1966.

[13] D. J. Newman and S. S. Bishton, Chem. Phys. Letters **1**, 616 (1968).

TABLE VIII. Analysis of Dy^{3+}:YAG parameters ($A_n^m\langle r^n\rangle$ and $\bar{A}_n$ are given in cm^{-1}).

	Experiment	No distortion	Angular distortion	Angular distortion (compensated)	Rolling-ligand model
$A_4^0\langle r^4\rangle$	−292	−288	−291	−293	−294
$A_4^2\langle r^4\rangle$	240	241	240	239	236
$A_4^4\langle r^4\rangle$	1046	1074	1056	1037	1033
$\bar{A}_4(R_1)$		118.7	123.3	121.9	96.2
$\bar{A}_4(R_2)$		71.3	65.3	73.7	96.0
t_4		9.3	11.6	9.2	0.0
$A_6^0\langle r^6\rangle$	41	50	41	42	41
$A_6^2\langle r^6\rangle$	−142	−96	−129	−151	−124
$A_6^4\langle r^6\rangle$	747	709	750	738	750
$A_6^6\langle r^6\rangle$	−77	−43	−59	−96	−48
$\bar{A}_6(R_1)$		29.8	28.3	28.5	46.2
$\bar{A}_6(R_2)$		23.5	24.0	20.3	17.0
t_6		4.3	3.0	6.2	4.3
$\Delta\theta_2$			−2.4°	−1.7°	
$\Delta\varphi_2$			+2.5°	8.8°	
R_1 (Å)	2.3030				2.0798
R_2 (Å)	2.4323				2.6228

TABLE IX. Analysis of Sm^{3+}:YGG parameters ($A_n^m\langle r^n\rangle$ and $\bar{A}_n$ are given in cm^{-1}).

	Experiment	No distortion	Angular distortion (two-parameter)	Angular distortion (one-parameter)
$A_4^0\langle r^4\rangle$	−231	−275	−235	−267
$A_4^2\langle r^4\rangle$	46	40	35	18
$A_4^4\langle r^4\rangle$	1283	964	1261	1045
$\bar{A}_4(R_1)$		93.5	71.4	91.9
$\bar{A}_4(R_2)$		79.4	114.7	76.4
t_4		4.4	−12.7	4.9
$A_6^0\langle r^6\rangle$	56	55	71	51
$A_6^2\langle r^6\rangle$	−21	−103	−149	−177
$A_6^4\langle r^6\rangle$	949	947	846	941
$A_6^6\langle r^6\rangle$	88	−23	−17	−61
$\bar{A}_6(R_1)$		37.4	44.3	38.8
$\bar{A}_6(R_2)$		30.2	24.4	23.1
t_6		5.7	15.9	13.9
$\Delta\theta_2$			7.4°	
$\Delta\varphi_2$			−1.1°	9.0°

TABLE X. Analysis of Sm^{3+}:YAG parameters ($A_n^m\langle r^n\rangle$ and $\bar{A}_n$ are given in cm^{-1}).

	Experiment	No distortion	Angular distortion (two-parameter)	Angular distortion (one-parameter)
$A_4^0\langle r^4\rangle$	−246	−289	−264	−282
$A_4^2\langle r^4\rangle$	165	156	132	136
$A_4^4\langle r^4\rangle$	1375	1081	1282	1154
$\bar{A}_4(R_1)$		113.7	101.3	111.8
$\bar{A}_4(R_2)$		76.7	83.8	73.9
t_4		7.2	3.5	7.6
$A_6^0\langle r^6\rangle$	45	66	70	62
$A_6^2\langle r^6\rangle$	−164	−163	−252	−213
$A_6^4\langle r^6\rangle$	1038	928	869	935
$A_6^6\langle r^6\rangle$	42	−45	−96	−94
$\bar{A}_6(R_1)$		40.9	45.1	41.5
$A_6(R_2)$		29.3	19.9	24.0
t_6		6.1	15.0	10.1
$\Delta\theta_2$			3.4°	
$\Delta\varphi_2$			8.6°	8.6°

in view of the deviations between experimental and fitted parameters.

G. Yb^{3+}:YGG

We have analyzed the parameters given by Pearson *et al.*,[14] who employed *g* factors as well as the crystal-field energy levels in their derivation. The fits obtained (see Table XI) using the superposition approximation are very poor, especially to the parameters $A_6^2\langle r^6\rangle$ and $A_6^6\langle r^6\rangle$, which are noticeably larger than the corresponding parameters in any other substituted garnet. For this reason all the results derived for this system must be regarded as extremely unreliable, especially in the angular-distortion model.

TABLE XI. Analysis of Yb^{3+}:YGG parameters ($A_n^m\langle r^n\rangle$ and $\bar{A}_n$ are given in cm^{-1}).

	Experiment	No distortion	Angular distortion
$A_4^0\langle r^4\rangle$	-177 ± 8.3	-177	-171
$A_4^2\langle r^4\rangle$	221 ± 48	221	226
$A_4^4\langle r^4\rangle$	621 ± 42	618	658
$\bar{A}_4(R_1)$		71.8	71.0
$\bar{A}_4(R_2)$		40.2	37.7
t_4		15.5	16.9
$A_6^0\langle r^6\rangle$	49 ± 5.4	51	49
$A_6^2\langle r^6\rangle$	-462 ± 60	-261	-253
$A_6^4\langle r^6\rangle$	822 ± 45	835	859
$A_6^6\langle r^6\rangle$	-233 ± 42	40	-14
$\bar{A}_6(R_1)$		42.0	41.0
$\bar{A}_6(R_2)$		19.1	17.4
t_6		20.9	22.9
$\Delta\theta_2$			0.3°
$\Delta\varphi_2$			9.1°

An analysis has been carried out for the Yb^{3+}:YAG system, also using experimental parameters obtained by Pearson *et al.*[14] Similar difficulties are encountered.

H. Tm^{3+}:YAG and Eu^{3+}:YGG

The experimental parameters for these systems are taken from the work of Koningstein[5]; the analyses are given in Tables XII and XIII. It is clear that Koningstein's parameters are meant to be approximate (no values being given for $A_6^6\langle r^6\rangle$), so the fits obtained are reasonable except in the case of $A_4^4\langle r^4\rangle$.

VI. DISCUSSION

A comparison of the results obtained in all the angular-distortion analyses (without compensation for distant-ion electrostatic effects) is given in Table XIV. Mean values of the intrinsic parameters {i.e., $\frac{1}{2}[\bar{A}_n(R_1)+\bar{A}_n(R_2)]$} are given because these are found to be effectively independent of the fitting procedure used. The general tendency for $\bar{A}_6$ to decrease with increasing Z and for $\bar{A}_4$ to remain almost constant is in

[14] J. J. Pearson, G. F. Herrman, K. A. Wickersheim, and R. A. Buchanan, Phys. Rev. **159**, 251 (1967).

TABLE XII. Analysis of Tm^{3+}:YAG parameters ($A_n^m\langle r^n\rangle$ and $\bar{A}_n$ are given in cm^{-1}).

	Experiment	No distortion	Angular distortion
$A_4^0\langle r^4\rangle$	-170	-206	-202
$A_4^2\langle r^4\rangle$	140	133	114
$A_4^4\langle r^4\rangle$	1020	771	824
$\bar{A}_4(R_1)$		82.5	81.1
$\bar{A}_4(R_2)$		53.5	51.9
t_4		7.9	8.2
$A_6^0\langle r^6\rangle$	30	33	31
$A_6^2\langle r^6\rangle$	-115	-106	-109
$A_6^4\langle r^6\rangle$	475	459	469
$A_6^6\langle r^6\rangle$	...	-14	-46
$\bar{A}_6(R_1)$		21.5	21.0
$\bar{A}_6(R_2)$		13.5	12.0
t_6		8.5	10.3
$\Delta\theta_2$			...
$\Delta\varphi_2$			8.7°

TABLE XIII. Analysis of Eu^{3+}:YGG parameters ($A_n^m\langle r^n\rangle$ and $\bar{A}_n$ are given in cm^{-1}).

	Experiment	No distortion	Angular distortion
$A_4^0\langle r^4\rangle$	-170	-194	-188
$A_4^2\langle r^4\rangle$	100	97	86
$A_4^4\langle r^4\rangle$	850	678	730
$\bar{A}_4(R_1)$		69.9	68.9
$A_4(R_2)$		52.1	49.8
t_4		7.9	8.7
$A_6^0\langle r^6\rangle$	50	54	50
$A_6^2\langle r^6\rangle$	-150	-142	-149
$A_6^4\langle r^6\rangle$	945	928	944
$A_6^6\langle r^6\rangle$	...	-7.7	-74
$\bar{A}_6(R_1)$		38.9	37.4
$\bar{A}_6(R_2)$		27.8	24.2
t_6		8.9	11.6
$\Delta\theta_2$			...
$\Delta\varphi_2$			9.1°

TABLE XIV. Comparison of derived parameters.

	Mean $\bar{A}_4$ (cm^{-1})	Mean $\bar{A}_6$ (cm^{-1})	t_4	t_6	$\Delta\theta_2$	$\Delta\varphi_2$
Sm^{3+}:YGG	84.2	31.0	4.9	13.9	...	9.0°
Eu^{3+}:YGG	59.4	30.8	8.7	11.6	...	9.1°
Dy^{3+}:YGG	88.3	23.7	11.4	8.1	0.9°	5.7°
ErGG	73.0	20.9	11.3	12.6	−2.3°	9.1°
Er^{3+}:YGG	75.5	20.6	14.5	6.6	−0.2°	9.1°
Yb^{3+}:YGG	54.4	29.2	16.9	22.9	0.3°	9.1°
Sm^{3+}:YAG	92.9	32.8	7.6	10.1	...	8.6°
Dy^{3+}:YAG	94.3	26.2	11.6	3.0	−2.4°	2.5°
ErAG	81.3	21.9	10.1	7.4	−1.3°	8.8°
Tm^{3+}:YAG	66.5	16.5	8.2	10.3	...	8.7°
Er^{3+}:LuGG	77.1	21.4	9.5	11.1	−0.9°	7.9°

accord with the experimental data obtained for ions substituted into $LaCl_3$.[7]

A comparison of results for the various systems containing Er^{3+} shows that the mean value of $\bar{A}_6$ is almost constant, while small variations occur in the mean value of $\bar{A}_4$. This is consistent with the previously expressed idea[6] that the ligand distances tend to become similar for a given metal–ligand system in any environment. The small differences in $\bar{A}_4$ may be due to variations in the distant electrostatic contributions.

The intrinsic parameters $\bar{A}_n(R)$ for ions substituted into both YAG and YGG generally have the correct ordering of values with respect to the undistorted distances. This is not true, however, for the Er^{3+}:LuGG parameters, and a comparison of these with parameters obtained in other Er^{3+} systems leads to the conclusion that the *relative* inward motion of ligands is approximately 0.03 Å greater for the yttrium garnets than it is in lutetium garnet. No deduction can be made about the *absolute* motion in either material.

Ratios of intrinsic parameters $\bar{A}_4/\bar{A}_6$ are given in Table XV, the quoted errors referring to the maximum for both distances and in several fitting procedures. These quantities are of interest because they show great stability and compare well with the experimental results obtained for the same ions in $LaCl_3$, except in the case of Sm^{3+} which is anomalous in both materials. The value quoted for Yb^{3+}:YGG is also clearly anomalous. It appears that these ratios depend on the rare-earth ion rather than the host crystal. Varying values of $\bar{A}_4/\bar{A}_6$ may be attributed to changes in the proportion of the electrostatic contribution to $\bar{A}_4$.

An attempt has recently been made by one of the authors to derive intrinsic parameters from experimental splittings in Er_2O_3.[15] This material has two distinct rare-earth sites with a mean metal–ligand distance of about 2.3 Å. The parameter $\bar{A}_6$ is found to be fairly independent of the assumptions made about the experimental data and has the value 21 ± 1 cm^{-1}. $\bar{A}_4$, however, takes alternative values according to the apparent power law required to fit the data, viz.,

$$t_4<7 \quad \text{gives} \quad \bar{A}_4=73\ \text{cm}^{-1},$$

$$t_4>13 \quad \text{gives} \quad \bar{A}_4=55\ \text{cm}^{-1}.$$

The values of $\bar{A}_6$ and the first value of $\bar{A}_4$ are in close accord with the mean values of these parameters that have been found for Er^{3+} in the garnets (see Table XIV).

It is also of interest to compare the magnitudes of the intrinsic parameters with those obtained theoretically, although such results are not yet available for O^{2-} ligands. They have a very similar range of values to those calculated for the system Pr^{3+}–F^- using an LCAO–MO model, giving, for example, at $R=2.416$ Å, $\bar{A}_4=98.3$ cm^{-1}, $\bar{A}_6=31.0$ cm^{-1}, and at $R=2.643$ Å, $\bar{A}_4=61.7$ cm^{-1}, $\bar{A}_6=19.6$ cm^{-1}. Such a comparison seems justified because the F^- and O^{2-} ligands are very

TABLE XV. Comparison of parameter ratios $\bar{A}_4/\bar{A}_6$ for ions substituted into garnets and $LaCl_3$.

Ion	Host: YGG[a]	YAG[a]	LuGG	$LaCl_3$
Sm^{3+}	2.64±0.06	2.80±0.09		1.38
Eu^{3+}	1.87±0.05			1.97
Dy^{3+}	3.66±0.08	3.54±0.07		3.47
Er^{3+}	3.50±0.16	3.63±0.09	3.51±0.11	3.51
Tm^{3+}		3.97±0.08		
Yb^{3+}	1.91±0.05			

[a] Numbers quoted in these columns also refer to pure salts.

[15] G. E. Stedman, Ph.D. thesis, University of London, 1968.

similar in size, and (hence) the metal–ligand distances are also comparable.

A recent determination of Yb^{3+}–O^{2-} exchange integrals by Pearson and Hermann[16] allows us to calculate the exchange contributions to the intrinsic parameters for the metal–ligand separation 2.35 Å. Using Eq. (3.1) of Ref. 17, we obtain

$$A_4^{ex} = -6.5 \text{ cm}^{-1}, \qquad A_6^{ex} = -2.6 \text{ cm}^{-1}.$$

These negative contributions of about 10% of the total parameters accord very well with the value of the exchange term adopted in our previous LCAO–MO calculations.[17]

It is apparent from Table XIV that there is a considerable scatter in our derived power laws. Even the most accurate fits to experimental data give power laws which deviate considerably from the mean values, presumably because of the artificial constraints imposed by a two-parameter angular distortion. Our results for t_n are therefore best discussed in terms of various average values.

Consider first the YGG power laws, omitting the case of Yb^{3+} because of the poor fit to experimental data and the large deviation of the t_n from their average values in this system. Taking straight averages for the other ions, we obtain

$$\text{Av } t_4 = 10.2 \pm 2.7, \qquad \text{Av } t_6 = 10.6 \pm 2.4,$$

where the quoted errors are mean deviations. A much better result may be obtained if we note that a large value of t_4 generally implies a small value of t_6, and vice versa. On the expectation that t_4 is approximately equal to t_6, we evaluate

$$\text{Av } \tfrac{1}{2}(t_4 + t_6) = 10.4 \pm 0.7,$$

which has considerably less error.

A similar analysis of the YAG results is spoilt by the low value t_6 derived for Dy^{3+}:YAG. Omitting this, we obtain averages which are reasonably consistent with those given above:

$$\text{Av } t_4 = 8.6 \pm 1.0, \qquad \text{Av } t_6 = 9.3 \pm 1.2,$$

$$\text{Av } \tfrac{1}{2}(t_4 + t_6) = 9.0 \pm 0.2.$$

There is a definite tendency for the t_n to be lower for YAG than they are for YGG, although the mean ligand distances in the undistorted host crystals are very similar in both cases. This may be explained on the basis that $R_1 - R_2$ is greater in YAG than in YGG, whereas, round a substituted ion, this difference would become less pronounced. Our derivation of t_n using the angular-distortion model does, of course, make no allowance for possible changes in the ligand distances.

[16] J. J. Pearson and G. F. Hermann, J. Appl. Phys. **40**, 1142 (1969).

[17] M. M. Ellis and D. J. Newman, J. Chem. Phys. **47**, 1986 (1967).

The separate averages for YAG and YGG (as well as the Er^{3+}:LuGG result) indicate that t_4 and t_6 are very nearly equal. This is in accord with theoretical results that have been obtained[7,9] for the systems Pr^{3+}–Cl^- and Pr^{3+}–F^- (see Sec. IV). It will be noticed that the values of t_n for O^{2-} ligands are closer to those predicted theoretically for Cl^- than to the F^- values. This is opposite to what might be expected purely on the basis of ligand size. However, it is not yet known what factors are primarily responsible for causing the differences between the F^- and Cl^- results; certainly they are not due to differences in the degree of covalency.

Finally, we remark on the very consistent nature of the rather large angular distortions that have been derived (see Table XIV). In all cases there is a tendency for the angle φ_2 to become 90°, and this gives a considerable improvement to the parameter fits, especially in the case of the $n=6$ parameters which are very sensitive to angular changes. Our only reservations arise from the magnitude of these predicted distortions, which are considerably greater than those that have been obtained for ions substituted into $LaCl_3$.

The relation of the φ_2 distortion to experimental parameters can be readily understood in terms of the need to increase the magnitudes of $A_4^4\langle r^4\rangle$, $A_6^2\langle r^6\rangle$, and $A_6^6\langle r^6\rangle$ simultaneously relative to the values obtained in the undistorted fits. For example, $A_n^n\langle r^n\rangle \sim \sin^n\theta \cos n\varphi$, and it is easily verified that changes in φ_2, but not φ_1, produce simultaneous changes in $A_4^4\langle r^4\rangle$ and $A_6^6\langle r^6\rangle$ of the correct relative sign to improve the fit to experimental data.

Because of the large values of $\Delta\varphi_2$ found in this analysis, a further check has been made to ensure that this result is physically reasonable by performing angular distortion analyses in which $\Delta\varphi_2 = 0$ and all three other angles φ_1, θ_1, θ_2 are varied. These analyses provide fits to the data in which the total angular distortion is again of the order of 9°, although the actual angles vary widely from case to case. We are led to reject these results mainly on the basis of the unphysical and mutually inconsistent values of t_4 and t_6 that are derived; typically $t_4 = 22$, $t_6 = -9$.

VII. CONCLUSION

Our main result is that the superposition approximation gives a fit to garnet crystal-field parameters which is generally as good as their expected accuracy. The determination of definite limits to the accuracy of this approximation requires even more accurate experimental data than is at present available. As the inclusion of distant ion electrostatic contributions generally spoils parameter fits, it may be the case that even the $n=4$ parameters are effectively screened.

Predicted intrinsic parameters and local angular distortions are insensitive to the method of analysis, while the derived power laws show a considerable unexplained scatter. Detailed comparison with theory is not yet possible, but the results obtained fit in well with the theoretical picture of the rare-earth crystal field that has been built up in recent years.[17,18] Certainly they are consistent with the idea that overlap and covalency give the dominant contributions to both the $n=4$ and $n=6$ parameters.

[18] M. M. Ellis and D. J. Newman, J. Chem. Phys. **49**, 4037 (1968).

ACKNOWLEDGMENTS

The authors are indebted to Professor H. M. Crosswhite and Dr. E. Orlich for supplying unpublished data, corrections, and comments. We also wish to thank Miss S. S. Bishton for reading an earlier draft of the manuscript. One of us (G.E.S.) is grateful to the Drapers' Company for a Research Studentship.

Appendix 2B: Electronic Polarisabilities of Ions in Alkali Halide Crystals†

S. S. JASWAL and T. P. SHARMA
Behlen Laboratory of Physics, University of Nebraska, Lincoln, Nebraska 68508, U.S.A.

(*Received* 28 *June* 1972)

Abstract—Tessman, Kahn and Shockley calculated the electronic polarisabilities of ions in alkali halide crystals using the long wavelength limiting values of the visible light dielectric constants. We have recalculated these widely used polarisabilities using the more accurate room-temperature dielectric constant data of Lowndes and Martin and a better minimisation procedure of Pirenne and Kartheuser. We have also calculated for the first time the low temperature values of these polarisabilities. The computed values of the polarisability in $Å^3$ are Li^+ 0·029, Na^+ 0·285, K^+ 1·149, Rb^+ 1·707, Cs^+ 2·789, F^- 0·876, Cl^- 3·005, Br^- 4·168, I^- 6·294 at 300°K and Li^+ 0·029, Na^+ 0·290, K^+ 1·133, Rb^+ 1·679, Cs^+ 2·743, F^- 0·858, Cl^- 2·947, Br^- 4·091, I^- 6·116 at 4°K. The relative standard deviations for all the alkali halides are 1·20 and 1·43 per cent at 300°K and 4°K respectively justifying the additive nature of the individual ion polarisabilities.

TESSMAN, Kahn and Shockley [1] (TKS) calculated the electronic polarisabilities of ions in alkali halide crystals using the long wavelength limiting values (ϵ_∞) of the visible light dielectric constants. These polarisabilities are widely used in studying the lattice properties of these crystals. We have recalculated the polarisabilities for the following reasons:

(i) Recently Lowndes and Martin [2] have provided the more accurate room-temperature and mostly new low temperature values of ϵ_∞.
(ii) Pirenne and Kartheuser [3] (PK) have given a better minimisation procedure to derive the individual ion polarisabilities.

The polarisability per unit cell, α_{ij}, is derived from ϵ_∞ using the standard Clausius–Mosotti relation

$$4\pi\alpha_{ij}/3 = v_a(\epsilon_\infty - 1)/(\epsilon_\infty + 2) \qquad (1)$$

where i and j refer to the constituent ions of the crystal and v_a is the volume per unit cell. Assuming the additive nature of the individual ion polarisabilities α_i,

$$\alpha_{ij} = \alpha_i + \alpha_j. \qquad (2)$$

To get the individual ion polarisabilities, TKS minimised

$$S = \sum_{i \neq j} (\Delta\alpha_{ij}^2), \qquad (3)$$

where $\Delta\alpha_{ij} = \alpha_{ij} - \alpha_i - \alpha_j$ and the summation is over all the alkali halides. They assumed the polarisability of Li^+ to be 0·029 which is the theoretical values due to Pauling. This procedure gives relative deviations $\Delta\alpha_{ij}/\alpha_{ij}$ as high as 27 per cent corresponding to wavelength $\lambda = 5893$ Å as shown by PK and 13 per cent for $\lambda = \infty$ as shown in Table 3. PK suggested an improvement over this procedure where one minimises

$$\sigma = \left[N^{-1} \sum_{i \neq j} \{\Delta\alpha_{ij}^2/\alpha_{ij}^2\}\right]^{1/2} \qquad (4)$$

where N is the number of alkali halides. We follow this procedure to find individual ion polarisabilities assuming the polarisability of Li^+ to be 0·029 Å [3].

The input data are given in Table 1 where we have included the TKS α_{ij} for comparison. We assume ϵ_∞ at 4°K (300°K) to be the same as at 2°K (290°K). The individual ion polarisabilities computed by minimising (4) are given in Table 2 along with the TKS values. The relative polarisability deviations $\Delta\alpha_{ij}/\alpha_{ij}$ corresponding to 4 and 300°K are given in

†Reprinted from *J. Phys. Chem. Solids* 1973 **34** 509–11.

Table 1. Input data and computed values of α_{ij}

	4°K			300°K			TKS values
	v_a[a] (Å³)	ϵ_∞[b]	α_{ij} (Å³)	v_a[c] (Å³)	ϵ_∞[b]	α_{ij} (Å³)	of α_{ij} (Å³)
LiF	15·90	1·93	0·898	16·20	1·93	0·915	0·909
LiCl	32·73	2·79	2·920	33·74	2·75	2·968	2·903
LiBr	39·94	3·22	4·055	41·62	3·16	4·159	4·138
LiI	51·40	3·89	6·020	54·00	3·80	6·223	6·226
NaF	24·17	1·75	1·154	24·65	1·74	1·164	1·163
NaCl	43·39	2·35	3·214	44·86	2·33	3·289	3·263
NaBr	51·55	2·64	4·350	53·28	2·60	4·424	4·388
NaI	65·17	3·08	6·370	68·27	3·01	6·539	6·264
KF	37·13	1·86	1·975	38·21	1·85	2·014	2·008
KCl	60·51	2·20	4·127	62·30	2·17	4·173	4·173
KBr	69·42	2·39	5·247	71·87	2·36	5·352	5·293
KI	84·94	2·68	7·280	88·17	2·65	7·469	7·388
RbF	43·39	1·94	2·471	44·85	1·93	2·534	2·572
RbCl	69·23	2·20	4·720	71·25	2·18	4·802	4·712
RbBr	79·30	2·36	5·905	80·49	2·34	5·933	5·920
RbI	95·51	2·61	7·963	98·94	2·58	8·148	8·093
CsF	53·07[d]	2·17	3·555	54·21	2·16	3·609	3·674
CsCl	67·32	2·67	5·747	70·08	2·63	5·890	5·829
CsBr	76·01	2·83	6·875	78·73	2·78	6·999	7·020
CsI	91·25	3·09	8·944	95·23	3·02	9·148	9·119

[a] GHATE P. B., *Phys. Rev.* **139**, A1666 (1965).
[b] Ref. [2].
[c] WYCKOFF R. W. G., *Crystal Structures*, Vol. I, Wiley, New York (1964).
[d] BORN M. and HUANG K., *Dynamical Theory of Crystal Lattices*, Clarendon Press, Oxford (1954).

Table 2. Alkali and halogen ion polarisabilities in

	Li^+	Na^+	K^+	Rb^+	Cs^+	F^-	Cl^-	Br^-	I^-
Present Values 4°K	0·029	0·290	1·133	1·679	2·743	0·858	2·947	4·091	6·116
Present Values 300°K	0·029	0·285	1·149	1·707	2·789	0·876	3·005	4·168	6·294
TKS Values	0·029	0·255	1·201	1·797	3·137	0·759	2·974	4·130	6·199

Table 3. Relative polarisability deviations $\Delta\alpha_{ij}/\alpha_{ij}$ for TKS values

	Li	Na	I	Rb	Cs
F	13·31%	12·73%	2·34%	−1·10%	−8·10%
Cl	−3·44%	1·10%	−0·07%	−1·27%	2·03%
Br	−3·78%	0·04%	−0·68%	−0·15%	0·84%
I	−0·04%	−3·01%	−0·12%	1·18%	1·09%

Table 4. Relative polarisability deviations $\Delta\alpha_{ij}/\alpha_{ij}$ for present values for 4°K

	Li	Na	K	Rb	Cs
F	1·24%	0·55%	−0·82%	−2·67%	−1·30%
Cl	−1·92%	−0·71%	1·13%	2·03%	0·99%
Br	−1·61%	−0·71%	0·44%	2·29%	0·60%
I	−2·07%	−0·56%	0·42%	2·11%	0·96%

Table 5. Relative polarisability deviations $\Delta\alpha_{ij}/\alpha_{ij}$ for present values at 300°K

	Li	Na	K	Rb	Cs
F	1·12%	0·29%	−0·55%	−1·95%	−1·57%
Cl	−2·25%	−0·03%	0·44%	1·86%	1·61%
Br	−0·91%	−0·66%	0·65%	0·97%	0·59%
I	−1·59%	−0·61%	0·35%	1·81%	0·71%

Tables 4 and 5 respectively. The values of σ for 4 and 300°K are 1·20 and 1·43 per cent respectively.

Using a better minimisation procedure we have updated the widely used TKS values of the alkali and halogen ion polarisaibilities with the more accurate room-temperature data and have computed for the first time the 4°K values of these polarisaibilities. Except for Na^+ the low temperature values are slightly lower than the room temperature values (Table 2). Whereas the maximum value of the relative polarisability deviation in TKS calculations is 13·31 per cent (Table 3), the corresponding value here is only −2·67 per cent (Tables 4 and 5). The present values of the polarisability for Na^+, Cs^+ and F^- are appreciably different from those of TKS (Table 2). Finally the assumption of the additive nature of individual ion polarisabilities is well justified by the very low values of σ (1·20 and 1·43 per cent).

REFERENCES

1. TESSMAN J. R., KAHN A. H. and SHOCKLEY W., *Phys. Rev.* **92**, 890 (1953).
2. LOWNDES R. P. and MARTIN D. H., *Proc. R. Soc. Lond.* **A308**, 473 (1969).
3. PIRENNE J. and KARTHEUSER E., *Physica* **30**, 2005 (1964).

Chapter 3

The Finite Difference Method

3.1 Introduction

In the finite difference method the continuum of possible values of an independent variable is abandoned in favour of a discrete set of its values and corresponding associated sets of values of functions of the variable. Further approximation of continuum values of a function may be obtained either by decreasing the scale of the discretisation, or by interpolation on the discrete values, as described in Chapter 1. The method is of very wide application in the numerical solution of differential equations, both ordinary and partial. We shall illustrate it with reference to some examples of the latter. Implementation of the method consists of two steps; (i) the replacement of the original equations by their finite difference form, and (ii) the solution of these equations for the discrete values of the function. For step (i) we require the finite difference expressions to replace the derivatives of a function, and these we now derive.

Approximate expressions for the first and second derivatives of a function at a point x in terms of the values of the function at any two other points, say $x-\eta$ and $x+\zeta$, may be found by writing the Taylor expansion for the function at these two points:

$$f(x-\eta) = f(x) - \eta f'(x) + \frac{\eta^2}{2} f''(x) - \frac{\eta^3}{3!} f'''(x) + \frac{\eta^4}{4!} f''''(x) - \cdots$$

$$f(x+\zeta) = f(x) + \zeta f'(x) + \frac{\zeta^2}{2} f''(x) + \frac{\zeta^3}{3!} f'''(x) + \frac{\zeta^4}{4!} f''''(x) + \cdots$$

Subtracting these expressions and ignoring quadratic and higher terms gives

$$f'(x) \simeq \frac{f(x+\zeta) - f(x-\eta)}{\eta + \zeta}. \tag{3.1}$$

Ignoring cubic terms, adding ζ times the first expression to η times the second gives

$$f''(x) \simeq \frac{2}{(\zeta\eta^2 + \eta\zeta^2)} [\zeta f(x-\eta) - (\eta+\zeta) f(x) + \eta f(x+\zeta)]. \tag{3.2}$$

If $\eta = \zeta = h$ this reduces to

$$f''(x) \simeq \frac{f(x-h) - 2f(x) + f(x+h)}{h^2}. \tag{3.3}$$

The first term which was ignored in using this expression is $-2h^2 f''''(x)/4!$, i.e. the truncation error involved is $O(h^2)$.

This procedure can also be extended to partial derivatives. For example, if $f = f(x, y)$, its Laplacian at a point 0, $\partial^2 f/\partial x^2 + \partial^2 f/\partial y^2$, may be written in terms of values of the function at adjacent points, as indicated in figure 3.1:

$$\nabla^2 f \simeq \frac{(f_1 + f_3) + \lambda^2(f_2 + f_4) - 2(\lambda^2 + 1)f_0}{h^2} - \frac{2h^2}{4!}\left(\frac{\partial^4 f}{\partial x^4} + \frac{1}{\lambda^2}\frac{\partial^4 f}{\partial y^4}\right) \tag{3.4}$$

where $\lambda \equiv h/k$.

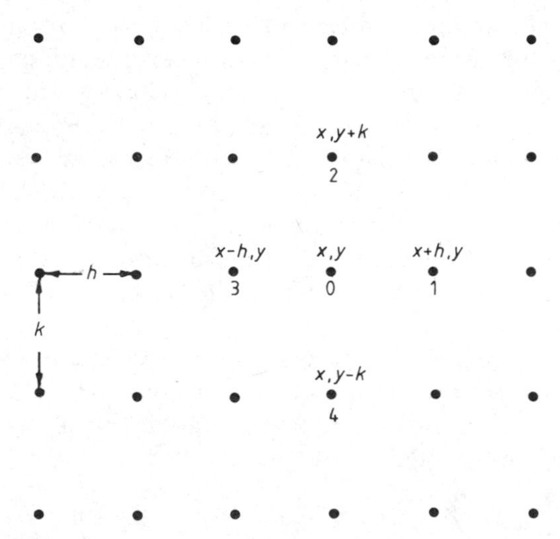

Figure 3.1 Using a rectangular mesh of unit size $h \times k$, the Laplacian of a function $f(x, y)$ at the point 0 may be approximated by expression (3.4).

3.2 Finite Difference Method and Partial Differential Equations

The replacement of a differential equation by equations for discrete values is a straightforward matter. However, the subsequent solution of the resulting difference equations is very much dependent on the nature of the original differential equation. There is a very large variety of such equations occurring in physics, further augmented by how the boundary conditions are specified, so for illustrative purposes we will have to restrict ourselves to a limited number of examples. It is necessary to distinguish between *initial value problems* where the function and some of its derivatives are specified on part of the boundary, often occurring in cases where one of the independent variables is the time t, and *boundary value problems* where information is specified on the boundary. We will discuss the latter case. More complete treatments are given in Ames (1977), Smith (1978) and Lapidus and Pinder (1982). The equation may be written formally in terms of a differential operator L as

$$L\Phi = \rho \tag{3.5}$$

and the boundary conditions expressed in the form

$$\alpha(s)\frac{\partial\Phi}{\partial n} + \beta(s)\Phi = \gamma(s) \tag{3.6}$$

where s denotes coordinates on the boundary, $\partial\Phi/\partial n$ is the derivative normal to the boundary, and α, β and γ are specified functions. Depending on these functions the problems are given different names;

Dirichlet problem $\quad \alpha = 0, \beta \neq 0 \rightarrow \Phi \quad$ specified on the boundary

Neumann problem $\quad \alpha \neq 0, \beta = 0 \rightarrow \dfrac{\partial\Phi}{\partial n} \quad$ specified on the boundary

Mixed problem $\quad \alpha \neq 0, \beta \neq 0 \rightarrow \dfrac{\partial\Phi}{\partial n} + \mu(s)\Phi = \nu(s)$ on the boundary.

An important family of equations results from the Sturm–Liouville operator

$$L \equiv -\operatorname{div}(p \operatorname{grad}) + g \tag{3.7}$$

where p and g are specified functions; a particular case of this gives Poisson's equation. Equations for a potential are basic to many simulation experiments, e.g. in fluid mechanics, plasma physics and astrophysics. In the latter case the evolution of a galaxy may be simulated by following the motion of a system of N bodies interacting among themselves through gravitation. The way this is done is to calculate the force on each body due to all the others and advance each of them under the force it experiences for a time Δt to find their new positions, continuing in this way to map the motion of the system. The interparticle forces may be evaluated directly from the N^2 pairs of interactions but for large N this becomes prohibitively slow and it is usual to derive them from the potential, the equation for which, Poisson's equation, must be solved after every interval Δt. An almost identical procedure can be used to study the behaviour of a plasma of electrical charges under different confinement conditions, which we briefly describe in §3.6. Perhaps the most important class of equations in this category is that describing the diffusion of neutrons in a nuclear reactor assembly, the solution of which determines under what conditions the assembly will become critical and many of the methods we describe were developed in the context of this problem.

We consider the operator (3.7) with p a constant and, for ease of exposition, restrict ourselves to the two-dimensional case, i.e. (writing $f \equiv g/p$, $q \equiv \rho/p$) the equation is

$$\frac{\partial^2 \Phi}{\partial x^2} + \frac{\partial^2 \Phi}{\partial y^2} + f(x, y)\Phi = q(x, y). \tag{3.8}$$

Its solution will be sought in a rectangular region XY, see figure 3.2, with Φ specified everywhere on the boundary, where we denote its values by F_i, the Dirichlet problem.

We cover the rectangle with a grid of $(n+1)(m+1)$ points, with spacing $h = X/n$, $k = Y/m$. Preferably, n and m should be to the power two. At any interior point of the mesh, labelled 0, equation (3.4) is used to approximate equation (3.8) at that point by ($\lambda \equiv h/k$)

$$\phi_{i+1,j} + \phi_{i-1,j} + \lambda^2\phi_{i,j+1} + \lambda^2\phi_{i,j-1} - [2(\lambda^2+1) - h^2 f_{ij}]\phi_{ij} = h^2 q_{ij} \tag{3.9}$$

where f_{ij} and q_{ij} indicate these functions evaluated at the point 0. A sequence of $(n-1)(m-1)$ such equations can be written down, some of which will involve one or more of the known values F_i, and they must be solved for the $(n-1)(m-1)$ unknown values of the potential. If we label each unknown potential with an index, e.g. in the order shown in figure 3.2, and consider the array ϕ of such values, we can write the system of $(n-1)(m-1)$ equations as

$$\mathbf{M}\phi = \boldsymbol{S} \tag{3.10}$$

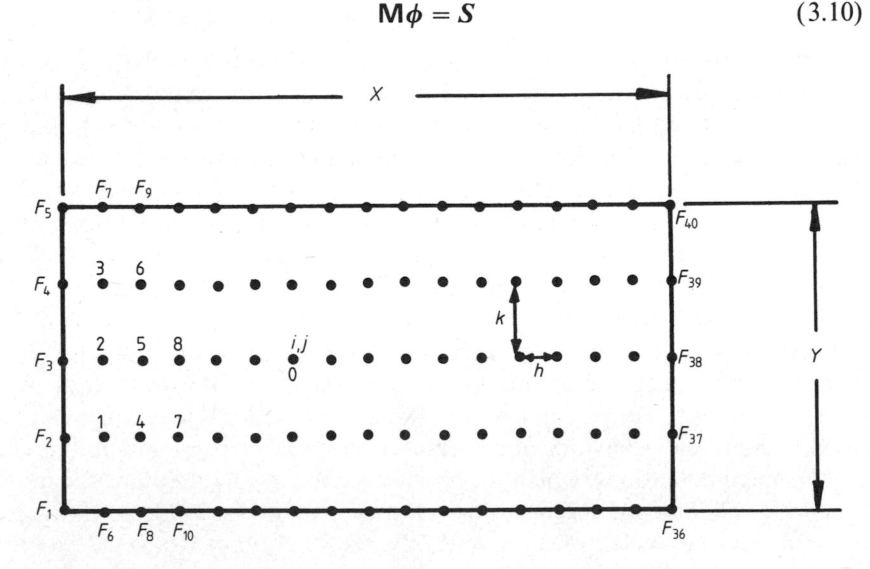

Figure 3.2 A 17 × 5 mesh for solving equation (3.8) in a rectangular region. The values F represent values of the solution known on the boundary.

where $\boldsymbol{S}$ is a column vector of $N = (n-1)(m-1)$ known quantities. To see what this equation looks like in detail let us take $f = 0$, i.e. Poisson's equation, and take $h = k$, i.e. $\lambda = 1$. Equation (3.9) written for the first five unknown values of the potential will be

$$
\begin{aligned}
\phi_1 - \tfrac{1}{4}\phi_2 \quad - \tfrac{1}{4}\phi_4 \quad &= -\tfrac{1}{4}(h^2 q_1 - F_2 - F_6)\\
-\tfrac{1}{4}\phi_1 + \phi_2 - \tfrac{1}{4}\phi_3 \quad - \tfrac{1}{4}\phi_5 \quad &= -\tfrac{1}{4}(h^2 q_2 - F_3)\\
-\tfrac{1}{4}\phi_2 + \phi_3 \quad - \tfrac{1}{4}\phi_6 \quad &= -\tfrac{1}{4}(h^2 q_3 - F_4 - F_7)\\
-\tfrac{1}{4}\phi_1 \quad + \phi_4 - \tfrac{1}{4}\phi_5 \quad - \tfrac{1}{4}\phi_7 \quad &= -\tfrac{1}{4}(h^2 q_4 - F_8)\\
-\tfrac{1}{4}\phi_2 \quad - \tfrac{1}{4}\phi_4 + \phi_5 - \tfrac{1}{4}\phi_6 \quad - \tfrac{1}{4}\phi_8 &= -\tfrac{1}{4}h^2 q_5.
\end{aligned}
$$

Here we see the known elements of $\boldsymbol{S}$ on the right-hand side and observe that $\mathbf{M}$ is of the form:

$$
\mathbf{M} = \begin{bmatrix}
1 & -\frac{1}{4} & 0 & -\frac{1}{4} & 0 & 0 & 0 & 0 & \cdots\\
-\frac{1}{4} & 1 & -\frac{1}{4} & 0 & -\frac{1}{4} & 0 & 0 & 0 & \cdots\\
0 & -\frac{1}{4} & 1 & 0 & 0 & -\frac{1}{4} & 0 & 0 & \cdots\\
-\frac{1}{4} & 0 & 0 & 1 & -\frac{1}{4} & 0 & -\frac{1}{4} & 0 & \cdots\\
0 & -\frac{1}{4} & 0 & -\frac{1}{4} & 1 & -\frac{1}{4} & 0 & -\frac{1}{4} & \cdots\\
\vdots & \vdots & \vdots & \vdots & \vdots & \vdots & \vdots & \vdots &
\end{bmatrix}
\tag{3.11}
$$

$\mathbf{M}$ is a symmetric matrix and its non-zero elements, all real, lie on a small number of diagonal lines. The problem is solved by solving the set of linear equations in (3.10) and there are different ways of doing this. Our discussion will not encompass the case where $f \neq 0$, $q = 0$, and all the elements of $\boldsymbol{S}$ are zero. This is an eigenvalue problem and will be treated in later chapters.

3.3 Solution of Finite Difference Equations

The solution of the differential equation is now replaced by the solution of (3.10). In principle, this equation can be solved exactly. However, we note that such a solution is still not the same as the solution of the original equation (3.8), hence the formerly unexplained change in notation from Φ to ϕ in equation (3.9). The difference between them, i.e. the truncation error, is a result of approximating the derivatives by the finite difference expressions, and will, in general, only vanish as $h \to 0$. In practice, it will be necessary to solve (3.10) for more than one value of h to investigate this convergence (see Exercise 3.1 at the end of the chapter). Different methods exist for solving (3.10), we list three such methods below, one of which is described in detail.

In a *direct solution* the equations are systematically solved either by matrix inversion, $\boldsymbol{\phi} = \mathbf{M}^{-1}\boldsymbol{S}$, or otherwise. In practice, the dimensions of $\mathbf{M}$ are often very large and numerical instabilities may occur due to the round-off process in the computer. For certain problems, notably Poisson's equation, direct methods can be used and are very fast if the problem is reformulated in terms of finite Fourier transforms of the unknown potential. Fourier transforms are described in Chapter 1, and some details of the method are given in §3.6.

A second method of solution is by means of a *random walk*. If we label the nearest neighbour points to any point 0, as in figure 3.1, we can rewrite (3.9) in the form

$$\phi_0 = \sum_{j=1}^{4} W_{0j}\phi_j + \mu q_0. \tag{3.12}$$

We can interpret the quantity W_{0j} as representing the relative importance of the neighbouring point j in determining the value of ϕ at the point 0. We will return to this method of solution by random walks in Chapter 7.

The most widely employed approach to solving the system of difference equations is known as *relaxation*, which is now described. We can write (3.10) in the form

$$\boldsymbol{\phi} = \mathbf{J}_0\boldsymbol{\phi} + \boldsymbol{S} \tag{3.13}$$

where $\mathbf{J}_0 = \mathbf{I} - \mathbf{M}$, and $\mathbf{I}$ is the unit matrix. Equation (3.13) is the basis for an iterative, or relaxation, solution; given some estimate $\boldsymbol{\phi}^{(k-1)}$ for the potentials we can obtain an 'improved' estimate $\boldsymbol{\phi}^{(k)}$

$$\phi_i^{(k)} = \sum_j J_{0ij}\phi_j^{(k-1)} + S_i. \tag{3.14}$$

$\mathbf{J}_0$ is a real, symmetric, matrix known as the Jacobi iteration matrix. An important property of this matrix is its spectral radius, $\rho(\mathbf{J}_0)$, which is the largest of the absolute values of its eigenvalues.

If we start with an arbitrary array of values (guesses) $\boldsymbol{\phi}^{(0)}$ for the potentials, it is not obvious that successive applications of (3.14) will converge to the solution $\boldsymbol{\phi}$ of the difference equation. At any stage of the iteration we can write an array of residuals $\boldsymbol{e}^{(k)}$ where $e_i^{(k)} = \phi_i^{(k)} - \phi_i$ and the ϕ_i values satisfy (3.13), the initial error array being $e_i^{(0)} = \phi_i^{(0)} - \phi_i$. We then have

$$\begin{aligned} \boldsymbol{e}^{(k)} &= \boldsymbol{\phi}^{(k)} - \boldsymbol{\phi} \\ &= \mathbf{J}_0\boldsymbol{\phi}^{(k-1)} + \boldsymbol{S} - (\mathbf{J}_0\boldsymbol{\phi} + \boldsymbol{S}) \\ &= \mathbf{J}_0(\boldsymbol{\phi}^{(k-1)} - \boldsymbol{\phi}) = \mathbf{J}_0\boldsymbol{e}^{(k-1)} \end{aligned} \tag{3.15}$$

i.e. $$\boldsymbol{e}^{(k)} = \mathbf{J}_0^k\boldsymbol{e}^{(0)}.$$

Hence, if $\mathbf{J}_0^k$ converges to the null matrix as $k \to \infty$, the iterative solution converges, independent of the choice of $\boldsymbol{\phi}^{(0)}$. Mathematically, the condition that $\mathbf{J}_0^k \to 0$ is that its spectral radius should satisfy $\rho(\mathbf{J}_0) < 1$. The fact that in no row of $\mathbf{J}_0$ (see (3.11), remember $\mathbf{J}_0 = \mathbf{I} - \mathbf{M}$) does the sum of all its terms exceed unity, and in at least one row is less than unity, ensures this inequality.

Almost as important as the fact of convergence is the rate at which it occurs. We define the norms of the vector of residuals and of the iterative matrix as follows:

$$\|\boldsymbol{e}^{(k)}\| = \left(\sum_i^N e_i^{(k2)}\right)^{1/2}$$

$$\|\mathbf{J}_0\| = \max_i \left(\sum_j |J_{0ij}|\right).$$

The *spectral norm*, $\|\mathbf{J}_0\|$, represents the maximum factor by which a vector of unit length is amplified when operated on by $\mathbf{J}_0$. Equation (3.15) then leads to

$$\|\boldsymbol{e}^{(k)}\| \leqslant \|\mathbf{J}_0\| \, \|\boldsymbol{e}^{(k-1)}\|.$$

For sufficiently large k this can be replaced by (see Exercise 3.2)

$$\|\boldsymbol{e}^{(k)}\| \lesssim \rho(\mathbf{J}_0) \, \|\boldsymbol{e}^{(k-1)}\|. \tag{3.16}$$

Thus, the closer the spectral radius $\rho(\mathbf{J}_0)$ is to unity the slower will be the rate of convergence. For the case of Poisson's equation on an $(m-1)(n-1)$ rectangular mesh, i.e. the $\mathbf{J}_0$ associated with (3.11) above, it can be shown that this maximum eigenvalue of $\mathbf{J}_0$ is

$$\rho(\mathbf{J}_0) = \frac{1}{2}\left(\cos\left(\frac{\pi}{m}\right) + \cos\left(\frac{\pi}{n}\right)\right). \tag{3.17}$$

With $n = m = 64$ this gives a reduction factor for the magnitude of the error vector of only 0.9988 per iteration. In addition, we see that as the mesh is made finer $\rho(\mathbf{J}_0)$ approaches closer to unity, resulting in ever slower convergence. A method of accelerating convergence will be discussed in §3.5.

The Jacobi method of relaxation is implemented as follows. Starting with a trial solution $\phi_i^{(0)}$, we can scan through all the points to derive successive estimates; for example, in our simple problem above

$$\phi_i^{(k+1)} = \tfrac{1}{4}(\phi_{(i1)}^{(k)} + \phi_{i2}^{(k)} + \phi_{i3}^{(k)} + \phi_{i4}^{(k)} - h^2 q_i) \tag{3.18}$$

where ϕ_{i1} etc are the four values surrounding the point from the previous iteration, as in figure 3.1 above. Although we have defined $\boldsymbol{\phi}$ as a vector with $(m-1)(n-1)$ components, in programming practice it will be more convenient to consider it as an $(m-1) \times (n-1)$ array.

We can, however, improve on equation (3.18). While carrying out the scan for the $(k+1)$th estimate, say at point 5 in figure 3.2, two new values, i.e. $\phi_2^{(k+1)}$ and $\phi_4^{(k+1)}$, have already been established, so we can use these rather than the estimates from the kth scan in (3.18). Equation (3.18) now becomes

$$\phi_i^{(k+1)} = \tfrac{1}{4}(\phi_{i1}^{(k)} + \phi_{i2}^{(k)} + \phi_{i3}^{(k+1)} + \phi_{i4}^{(k+1)} - h^2 q_i). \tag{3.19}$$

This is known as the *Gauss–Seidel* method. It corresponds to writing $\mathbf{M} = \mathbf{I} - \mathbf{L} - \mathbf{U}$, where $\mathbf{L}(\mathbf{U})$ is a matrix which has all elements on or above (below) the diagonal equal to zero, and replacing (3.13) by

$$\boldsymbol{\phi}_{\mathrm{GS}}^{(k+1)} = \mathbf{L}\boldsymbol{\phi}^{(k+1)} + \mathbf{U}\boldsymbol{\phi}^{(k)} + \boldsymbol{S}. \tag{3.20}$$

For the Gauss–Seidel method, it can be shown that the error vector decreases asymptotically as

$$\|\boldsymbol{e}^{(k)}\| = \rho^2(\mathbf{J}_0)\,\|\boldsymbol{e}^{(k-1)}\|$$

although at early stages of the iteration the convergence may be much faster.

In carrying out iterative solutions of this kind, since we do not know the actual answer in advance, some way must be found to decide when to accept the iterated result. That is, we need a criterion for convergence. We would like a solution such that the fractional error anywhere does not exceed a specified value ε, typically ε is the order of 10^{-3}, i.e. at every point

$$e_i^{(k)} \equiv |\phi_i^{(k)} - \phi_i| < \varepsilon\phi_i. \tag{3.21}$$

Since, however, the ϕ_i are not known, we must work in terms of the rate of convergence, i.e. the displacements $\delta_i^{(k)} = |\phi_i^{(k)} - \phi_i^{(k-1)}|$. One possibility is to make all $\delta_i^{(k)} < \varepsilon\phi_i^{(k)}$. In some problems it might be that this condition is fulfilled over most of the region, except for a small part of it in which the convergence is much slower, in such a case a more suitable criterion might be

$$\frac{\|\boldsymbol{\delta}^{(k)}\|}{\|\boldsymbol{\phi}^{(k)}\|} < \varepsilon. \tag{3.22}$$

The condition (3.22) by no means guarantees that our desired criterion (3.21) holds, depending, as it does, very much on the specific problem, and it is advisable when using it to set ε smaller than the acceptable tolerance by a factor of 10 or more.

Alternative criteria can be devised, based on how well the $\phi_i^{(k)}$ values satisfy the difference equation rather than on their rate of convergence, but, in practice, they are difficult to evaluate. As we indicated above, the rate of convergence is slower when the number of points on the mesh is greater. The inability to guarantee convergence in the solution of the difference equations is the greatest weakness of the method of relaxation.

To further appreciate these methods it will be instructive to look at a practical example, but before doing that there are two minor aspects which require attention.

For certain specified boundaries it may not be possible to choose a grid which coincides with the boundary, especially if the latter is curved. We can, however, use equation (3.4) to write the difference equation for the point 0 in figure 3.3(a) in terms of the values at points 1′ and 2′ on the boundary, rather than the mesh point values at 1 and 2. For example, we approximate ∇^2 by

$$\nabla_0^2 \simeq \frac{2}{h^2}\left[\frac{\phi_{1'}}{\alpha(1+\alpha)} + \frac{\lambda^2\phi_{2'}}{\beta(1+\beta)} + \frac{\phi_3}{1+\alpha} + \frac{\lambda^2\phi_4}{1+\beta} - \left(\frac{1}{\alpha} + \frac{\lambda^2}{\beta}\right)\phi_0\right]. \quad (3.23)$$

The second modification is required when normal derivatives, $\partial\Phi/\partial n$, rather than values of the function, are given on the boundary—the Neumann condition. For the case where the direction of the normal coincides with the mesh axis, we can use a replacement similar to that used to obtain (3.23) to approximate the Laplacian at 0 in figure 3.3(b) to

$$\nabla_0^2 \simeq \frac{2}{h^2}\left[\frac{\lambda^2\phi_2}{2} + \frac{\phi_3}{2\alpha+1} + \frac{\lambda^2\phi_4}{2} + \left(\frac{h}{2\alpha+1}\right)\frac{\partial\Phi}{\partial n} - \left(\frac{1}{2\alpha+1} + \lambda^2\right)\phi_0\right]. \quad (3.24)$$

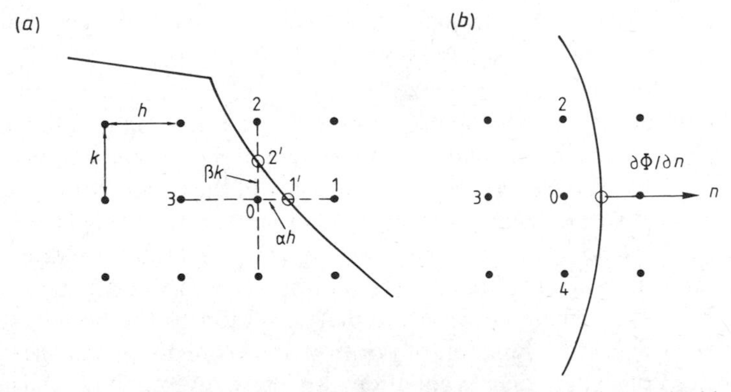

Figure 3.3 (a) Points 1′ and 2′ on the boundary, which does not coincide with the mesh, are used in approximating the derivatives at the point 0, see (3.23). (b) In the case of the normal derivative at the boundary being specified, expression (3.24) can be used in the replacement if the normal coincides with a mesh axis.

The case where the normal is not along a mesh axis is discussed in §§1–9 of the book by Ames (1977). These latter two formulae involve truncation errors of O(h), rather than O(h^2) as obtained by using the first term in (3.4).

3.4 An Example

Although numerical methods like those being discussed here are used for complicated problems, e.g. the relative permittivity as a function of position, for exposition purposes it is advantageous to focus on a simple case as follows. We wish to find the capacitance per unit length, normal to the page, of the channel formed by a pair of conductors, the left half of which is shown in figure 3.4. The conductors extend in a symmetric fashion an equal distance

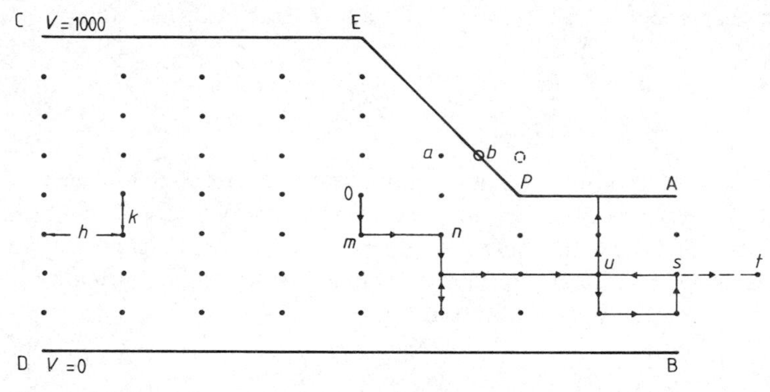

Figure 3.4 A mesh to determine the electrostatic potential between the capacitor plates. The convergence to a solution at the point 0 using different methods of iteration is shown in figure 3.5. The path 0mn... refers to a method of solution based on a random walk, described in §7.2.

to the right of the line AB and are separated by a medium of uniform relative permittivity. If one of the plates is at a potential V_0, the capacitance is given by $C = |Q|/V_0$, where Q is the charge on either of the conductors. $Q = \int \sigma \, \mathrm{d}A$, where σ is the surface charge density, which on a conductor is given in terms of the electric field by $\sigma = D = \varepsilon_r \varepsilon_0 E = -\varepsilon_r \varepsilon_0 \, \partial\Phi/\partial n$. Hence, we must find the potential in the region of one of the plates; to do this in practice, we must solve for Φ everywhere between the plates, subject to the boundary conditions. Since ε_r is not a function of position and there are no charges present, Φ satisfies $\nabla^2\Phi = 0$, Laplace's equation. An alternative method is to note that the energy (given in Chapter 6 by equation (6.2)) may also be written as $\frac{1}{2}CV_0^2$, so

$$C = \frac{\varepsilon_0 \varepsilon_r}{V_0^2} \int |\mathrm{grad}\ \Phi|^2 \, \mathrm{d}x \, \mathrm{d}y.$$

The problem is very similar to that we discussed above. For a point like *a* in figure 3.4, equation (3.23) must be used ($\alpha = \frac{1}{2}$, $\beta = 1$), while on the boundaries AB and CD we must use the normal derivative boundary conditions. On AB, because of symmetry, the field will be normal to the plates, i.e. $\partial\Phi/\partial n = 0$, while on CD we will have to use this same condition, which for the dimensions shown will be a reasonable approximation at all interior

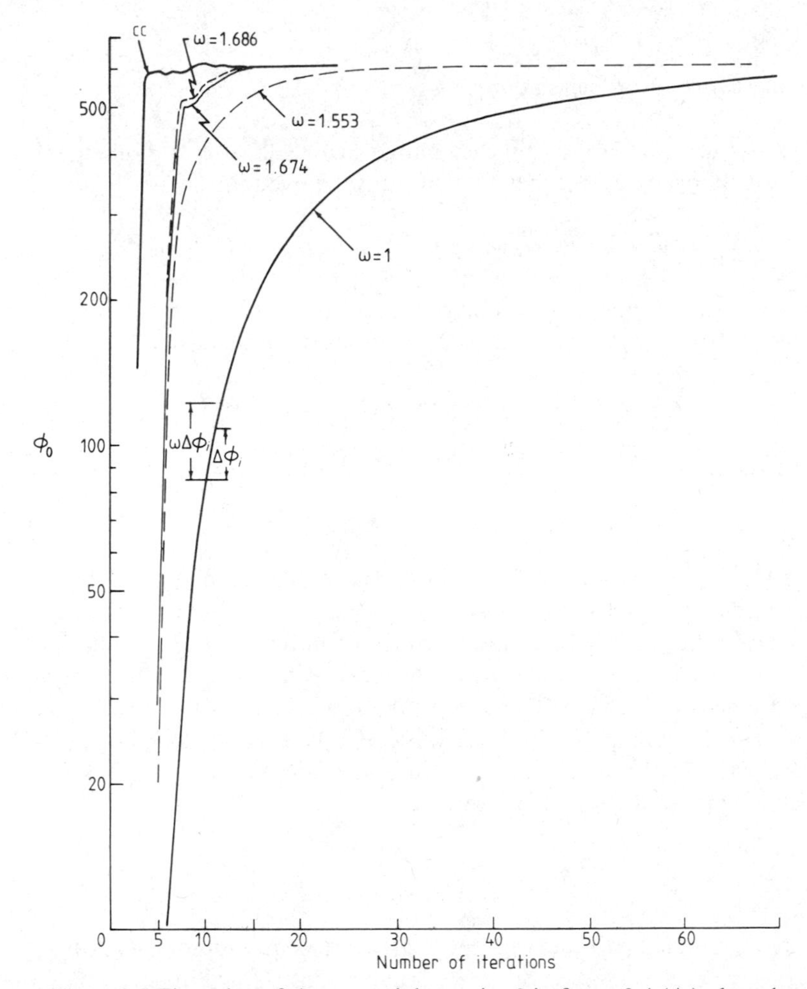

Figure 3.5 The value of the potential at point 0 in figure 3.4 (ϕ_0) plotted against the iteration number. The case $\omega = 1$ corresponds to the use of expression (3.20), Gauss–Seidel iteration. Other values of ω correspond to the use of successive over-relaxation as described in the text. The curve marked CC corresponds to use of the cyclic Chebyshev method described in Exercise 3.3 at the end of the chapter.

points along CD. With $V_0 = 1000$ and intitial values $\boldsymbol{\phi}^{(0)} = 0$ on a rectangular mesh 17×17, the rate at which the potential at the point 0 converges to its final value is shown by the curve marked $\omega = 1$ in figure 3.5. We see that after 70 iterations it is still more than 5% away from its asymptotic value. It would seem that we could get to this value more quickly if at each iteration we made a greater change in the potential, instead of changing ϕ_i by $\Delta\phi_i$, as in the figure, we change it by a larger amount, say $\omega\Delta\phi_i$. This is the method of over-relaxation.

3.5 Successive Over-Relaxation

Instead of using (3.20) to find the change $\Delta\phi = \phi_{i\,\mathrm{GS}}^{(k)} - \phi_i^{(k-1)}$ in each iteration, we change it by ω times this amount, i.e. we take

$$\boldsymbol{\phi}^{(k)} = \boldsymbol{\phi}^{(k-1)} + \omega(\boldsymbol{\phi}_{\mathrm{GS}}^{(k)} - \boldsymbol{\phi}^{(k-1)}) = \omega\boldsymbol{\phi}_{\mathrm{GS}}^{(k)} + (1-\omega)\boldsymbol{\phi}^{(k-1)} \tag{3.25}$$

where ω is a parameter to be determined, which will accelerate the convergence. Inserting from equation (3.20) we can write this as

$$\boldsymbol{\phi}^{(k)} = \mathbf{J}_\omega\boldsymbol{\phi}^{(k-1)} + \omega(\mathbf{I} - \omega\mathbf{L})^{-1}\boldsymbol{S} \tag{3.26}$$

where the iteration matrix is

$$\mathbf{J}_\omega = (\mathbf{I} - \omega\mathbf{L})^{-1}(\omega\mathbf{U} + (1-\omega)\mathbf{I}). \tag{3.27}$$

Equation (3.26) resembles (3.13) above and the development of the error $\boldsymbol{e}$ can be described in a similar way, i.e. $\boldsymbol{e}^{(k)} = \mathbf{J}_\omega\boldsymbol{e}^{(k-1)}$. In particular, at large k a relation like (3.16) may be written down in terms of $\rho(\mathbf{J}_\omega)$, the spectral radius of $\mathbf{J}_\omega$. Hence, a value of ω should be chosen which minimises the largest eigenvalue of $\mathbf{J}_\omega$. This is the method of successive over-relaxation (SOR). The dependence of this eigenvalue $\rho(\mathbf{J}_\omega)$ on ω is sketched in figure 3.6, after Carre (1961). It is seen to have a minimum in the range $1 < \omega < 2$ which is approached rather rapidly at

$$\omega_{\min} = \frac{2}{1 + (1 - \rho^2(\mathbf{J}_0))^{1/2}}. \tag{3.28}$$

The actual value of the asymptotic convergence factor $\rho(\mathbf{J}_\omega(\omega_{\min}))$ is $(\omega_{\min} - 1)$. For example, on a 2^6 square mesh for Poisson's equation, $(\omega_{\min} - 1) = 0.9065$, to be compared with $\rho(\mathbf{J}_0) = 0.9988$ found for the same mesh without over-relaxation in §3.3 above.

The above discussion has some drawbacks in practice. To find $\omega_{\min}$ we need to determine $\rho(\mathbf{J}_0)$: in general this is not easy. The value of $\omega_{\min}$ will lie between one and two, but since the convergence factor varies quite rapidly in the neighbourhood of $\omega_{\min}$, it is valuable to have a reasonably good

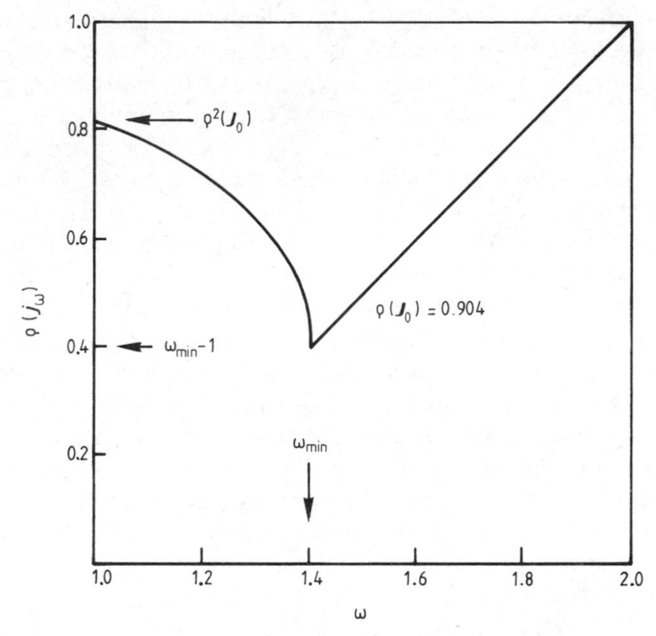

Figure 3.6 The spectral radius of the SOR matrix $\mathbf{J}_\omega$ as a function of ω for a problem for which the spectral radius of the Jacobi matrix is 0.904. Note the rapid variation in the neighbourhood of the minimum.

estimate of it. One suggestion is to take $\omega = 1$ for several iterations using the Gauss–Seidel method and find

$$\frac{\|\boldsymbol{\delta}^{(k)}\|}{\|\boldsymbol{\delta}^{(k-1)}\|}$$

which tends asymptotically to $\rho^2(\mathbf{J}_0)$. Alternatively, find the values m and n of the rectangular grid which most closely coincide with the grid used in the problem and use equation (3.17). With a value for ω available, the SOR method can be implemented directly using (3.25).

For our problem after 10 iterations, the ratio of the norms gives the estimate $\rho(\mathbf{J}_0) = 0.958$, $\omega = 1.553$, while (3.17) gives $\rho(\mathbf{J}_0) = 0.981$, $\omega = 1.674$ with $m = n = 16$, and the asymptotic ratio of the norms gives $\omega = 1.686$. The dramatic improvement in the rate of convergence at the point 0 is seen in figure 3.5. This suggests that even for a problem not of the Dirichlet type on a rectangle the estimate of $\rho(\mathbf{J}_0)$ using (3.17) may often be a good approximation to its true value. More sophisticated methods for estimating the spectral radius, and for carrying out the relaxation, are described in the literature; see, for example, Chapter 3 of Ames (1977). The curve marked CC in figure 3.5 is the result of using an odd/even cyclic Chebyshev method which is discussed in an exercise at the end of the chapter.

Whatever method is used, it is worthwhile to re-emphasise the distinction between the solution so obtained, ϕ, and the solution Φ to the original equation. In replacing the differential equation with a difference equation we employed (3.4), from which we see that if the fourth and higher derivatives of Φ vanish, i.e. Φ is a polynomial of degree three or less, the difference equation refers to the same function as the original equation. If this is not so, the solution of the difference equation, no matter how accurately determined, is not the same function which obeys the differential equation. Since the terms neglected in the replacement are $O(h^2)$, it would be expected that by taking a fine enough mesh we could get as close as desired to the solution of the differential equation, and indeed in an exercise at the end of the chapter we see that the fractional difference between the two functions is $O[(h/L)^2]$, where L is the dimension of the region. In general, the truncation error is greatest in regions where the function is most rapidly varying, e.g. in the region of the point P in figure 3.4. The discrepancy will be even greater if formulae like (3.23), for which the truncation error is $O(h)$, are used near such points. In the case where there is a singularity in the higher derivatives, the terms neglected in obtaining the difference equation will not be negligible for any finite h. In the neighbourhood of a point or line charge just such divergent behaviour occurs, so one must be careful not to replace conductors of small dimensions by lines or points. Methods exist for reducing the truncation errors in such cases, one of which is described in a project below.

Often, one does not have any idea of how the function behaves, so it seems at least necessary to repeat the calculation with a quadrupled number of points to get some idea of how the solution is converging; a procedure for improving the accuracy of the estimate based on repeated solutions of the difference equation is described in Exercise 3.1 at the end of the chapter. We have seen earlier that the convergence of the solution of the difference equation becomes slower as the size of the mesh is decreased. It is therefore necessary to compromise by using a mesh small enough to ensure that the solution of the difference equation is a reasonable approximation to the function sought and large enough to yield an accurate solution of the difference equation in a practical number of iterations. $N \sim 2^6$ divisions in each direction should prove a suitable compromise in many problems.

PROJECT 3A: DESIGN OF A PROPORTIONAL COUNTER

The configurations of equipotentials, and field lines, are important in the design of gas amplification counters, such as proportional counters. With cylindrical counters it is straightforward to design a counter in which the gain is independent of the striking location of the particle. However, it is often much more convenient to have counters of rectangular cross section, or counters containing many electrodes. This project is an introduction to the design of a simple counter, consisting of a single cylindrical anode, at a high potential, running along the axis of a square metal pipe which is grounded. The radius of the anode, a, will be much smaller than the side of the square L; take $L = 1$ cm and $a = 2^{-9}$ cm. Symmetry permits us to concentrate on just a quarter of the configuration, as shown in figure 3.7, in which Laplace's equation, $\nabla^2\Phi = 0$, must be solved subject to suitable boundary conditions. Although no singularity occurs in the field between the electrodes, it changes very rapidly in the immediate neighbourhood of the anode: this project illustrates one method by which the resulting truncation error, for a given mesh size, may be reduced; more detailed discussion is given in §24 of the book by Fox (1962).

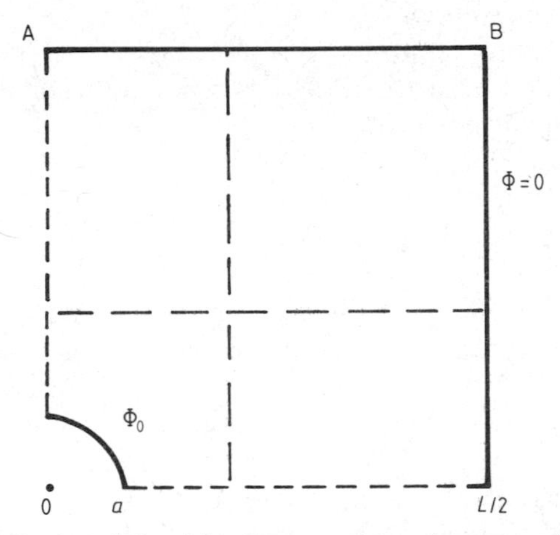

Figure 3.7 Region defined by the proportional counter; Project 3A.

Using a 17×17 mesh, set up the difference equation (note that (3.23) should be used for the mesh points nearest the anode) and solve it using SOR (or the cyclic Chebyshev method discussed in Exercise 3.3). If SOR is used, an estimate for the relaxation parameter should be determined as described in §3.5 above. You should compare the values found for ϕ along the central line with the known solution, some values of which, for $\Phi_0 = 100$, are given in table 3.1. Further values may be obtained using equations (4) and (11) in Tomitani (1972).

One way to reduce the truncation error is to employ a finer mesh in the region where the potential is varying most rapidly, i.e. in the neighbourhood

Table 3.1 Some values of the known solution of the differential equation with $\Phi_0 = 100$.

$18x$	1	2	3	4	5	6	7	8	9
$\Phi(x)$	33.6	23.3	17.3	13.0	9.7	6.9	4.4	2.2	0

of the anode. Investigate the effect of quadrupling the number of points in each direction in the innermost cell. Note that at the interface between meshes you will need to use equation (3.2) to rewrite the difference equation. For the same mesh size the truncation error can be greatly reduced by noticing that in two dimensions the function $f(r) = \ln r$, satisfies $\nabla^2 f(r) = 0$, so that a function Ψ, defined as $\Psi = \Phi + K\Phi_0 \ln(r/a)$, obeys $\nabla^2\Psi = 0$ in the same region, with appropriate boundary conditions. By a suitable choice of K, however, the extent of the variation of Ψ in the region may be much smaller than the variation of Φ. Choose a value of K which makes $\Psi = \Phi_0$ at the point A in figure 3.7, set up the difference equations and solve them as before to find ψ. Compare the resulting values of ϕ with the earlier values and with the known solution.

Further investigation can be pursued; for example, set $\Psi = \Phi_0$ at the point B, or use the methods discussed in Urbanczyk and Waligorski (1975) in the design of multi-anode counters.

*3.6 Direct Methods for Poisson's Equation

The speed of convergence to the solution of the difference equations may be very fast, as exemplified in figure 3.5 for the case of relaxation using the Chebyshev method. Even so, solutions of this kind may still pose a severe constraint on the speed with which the motions of a system of electrostatically interacting charges in a plasma may be simulated, and we use this problem as an illustration of a method of solution of the difference equation which does not involve relaxation. The evolution of the system is commonly analysed at discrete time steps as follows. A mesh of step size h is superimposed on the system and a charge density is assigned to each point of the mesh, corresponding to the actual charge in a cell of size $\pm h/2$ in each direction about the point at that time step. For this distribution, a solution for the potential is obtained using Poisson's equation and this potential is differentiated to find the force on the individual particles. The charges are then advanced, subject to these forces, for a time Δt, giving rise to the new charge distribution for the next time step, when Poisson's equation has to be solved again. Thus, the solution of the differential equation has become a frequent stage in a more extended calculation, and the time required to solve the potential equation should not be out of proportion to the time required for other steps of the simulation, otherwise it will severely constrain the size and particle numbers of the system which can be studied. The geometry and the boundary conditions in these problems can sometimes be assumed to have rather simple forms. If this is the case, a direct method of solving the difference equations, which avoids iteration, may be used.

The method involves finite Fourier series, discussed in §1.4 above, and the cyclic reduction (FACR), which is due to Hockney (1970). We will consider the very simple case of a rectangle of sides X and Y, each of which is a multiple of h ($X = nh$, $Y = mh$), with the potential vanishing everywhere on the boundary. At any value of x we can write $\Phi(x, y)$ as a Fourier series in y, as in (1.12), where, because of the boundary conditions, only the sine terms occur:

$$\Phi(x, y) = \sum_{k=1}^{\infty} B_k(x) \sin\left(\frac{2\pi yk}{Y}\right) \tag{3.29}$$

and

$$B_k(x) = \frac{2}{Y}\int_0^Y \Phi(x, y) \sin\left(\frac{2\pi ky}{Y}\right) \mathrm{d}y. \tag{3.30}$$

If we insert this into the equation

$$\frac{\partial^2 \Phi}{\partial x^2} + \frac{\partial^2 \Phi}{\partial y^2} = q(x, y)$$

and use the orthogonality properties of the trigonometrical functions, we obtain the equations satisfied by the coefficients $B_k(x)$. These are of the form

$$\frac{\mathrm{d}^2 B_k(x)}{\mathrm{d}x^2} - \frac{4\pi^2 k^2}{Y^2} B_k(x) = C_k(x) \qquad k = 1, \ldots, \infty \tag{3.31}$$

where C_k is the coefficient in the Fourier expansion of the (known) charge distribution:

$$C_k(x) = \frac{2}{Y}\int_0^Y q(x, y) \sin\left(\frac{2\pi ky}{Y}\right) \mathrm{d}y.$$

We replace (3.31) by a difference equation, i.e. we seek the coefficients at values of $x_i = ih$, and in conformity with our emphasis on the difference between the solutions of a differential equation and its difference equation, we denote them by $B_k(x_i) \to b_k^{(i)}$. They satisfy

$$b_k^{(i-1)} - \left(\frac{4\pi^2 k^2 h^2}{Y^2} + 2\right) b_k^{(i)} + b_k^{(i+1)} = h^2 C_k^{(i)} \qquad i = 1, \ldots, (n-1).$$

Coefficients $b_k^{(0)}$ and $b_k^{(n)}$, determined by the boundary conditions, are zero in this case. Rather than try to find the infinity of coefficients at each value of i, we content ourselves with the coefficients of a discrete Fourier expansion, (1.14), based on the $(m+1)$ values at $y_j = jh$, for which the coefficients are $b_k^{(i)}$ where $k = 1, \ldots, m/2$; that is

$$\phi(x_i, y_j) = \sum_{k=1}^{m/2} b_k^{(i)} \sin\left(\frac{2\pi jk}{m}\right). \tag{3.32}$$

They satisfy the equation

$$b_k^{(i-1)} - \lambda_k b_k^{(i)} + b_k^{(i+1)} = h^2 C_k^{(i)} \tag{3.33}$$

where $i = 1, \ldots, (n-1)$, $k = 1, \ldots, m/2$,

$$\lambda_k \equiv \left(\frac{4\pi^2 k^2 h^2}{Y^2} + 2\right)$$

and

$$C_k^{(i)} = \frac{2}{m} \sum_{j=0}^{m} q_{ij} \sin\left(\frac{2\pi kj}{m}\right). \tag{3.34}$$

The procedure for solution is thus to Fourier analyse the charge distribution, (3.34), solve (3.33) for the $m(n-1)/2$ coefficients $b_k^{(i)}$ and use them to carry out the Fourier synthesis (3.32).

At first sight, this seems to involve a great deal of work, however, we are aided by two circumstances. When m has an integral value, the methods of fast Fourier transform and synthesis, described in §1.4, may be used for the first and third steps, while a similar technique, *cyclic reduction*, which we describe below, may be used in the solution of equation (3.33).

For clarity of presentation, we consider a specific case with eight divisions in the x direction, i.e. $(n+1) = 9$. For any value of k, we can write equation (3.33) for three successive values of i

$$\begin{aligned}
b_k^{(i-2)} - \lambda_k b_k^{(i-1)} + b_k^{(i)} \qquad\qquad\qquad\qquad &= h^2 C_k^{(i-1)} \\
b_k^{(i-1)} - \lambda_k b_k^{(i)} + b_k^{(i+1)} \qquad\quad &= h^2 C_k^{(i)} \\
b_k^{(i)} - \lambda_k b_k^{(i+1)} + b_k^{(i+2)} &= h^2 C_k^{(i+1)}.
\end{aligned}$$

By adding λ_k times the second equation to the other two we find:

$$b_k^{(i-2)} - \lambda_k' b_k^{(i)} + b_k^{(i+2)} = h^2 C_k^{(i)\prime} \qquad i = 2, 4, 6 \tag{3.35}$$

where

$$\lambda_k' \equiv \lambda_k^2 - 2$$

$$C_k^{(i)\prime} \equiv C_k^{(i-1)} + \lambda_k C_k^{(i)} + C_k^{(i+1)}.$$

Equation (3.35) links values of the coefficients on alternate nodes in the x direction. We can continue in this way by writing three successive cases of (3.35) and manipulating them in the same manner. We get

$$b_k^{(i-4)} - \lambda_k'' b_k^{(i)} + b_k^{(i+4)} = h^2 C_k^{(i)\prime\prime} \qquad i = 4 \tag{3.36}$$

where

$$\lambda_k'' = (\lambda_k')^2 - 2$$

$$C_k^{(i)\prime\prime} = C_k^{(i-2)\prime} + \lambda_k' C_k^{(i)\prime} + C_k^{(i+2)\prime}.$$

In the present case we have gone as far as we can with this reduction, (3.36) expresses $b_k^{(4)}$ in terms of the known quantities $b_k^{(0)}$, $b_k^{(8)}$ and the coefficients (3.34). Knowing $b_k^{(4)}$ and setting $i = 2$ and 6 respectively in (3.35), we can find $b_k^{(2)}$ and $b_k^{(6)}$ in terms of known quantities. Finally, the values so far obtained can be used with equation (3.33) to determine all the remaining values $b_k^{(1)}$, $b_k^{(3)}$, $b_k^{(5)}$ and $b_k^{(7)}$.

The procedure in the general case is obvious; successive depths of reduction may be effected until one is left with a smaller number of equations, in the present case one, which can be readily solved. The method is only possible for particular types of boundary conditions, and in practice may be more complicated than in the above simple illustration—the case of periodic boundary conditions is discussed by Hockney (1970)—but in certain cases has been shown to be faster than iteration methods using relaxation.

3.7 Initial Value Problems

As mentioned in §3.1, a wide class of differential equations occurs in physics whose specification is different from that discussed so far. The most important represent unsteady heat or fluid flow, where time is one of the independent variables, technically they are known as parabolic equations. The region of the problem will extend in space and time; however, information on the value of the function or its derivatives will not be available initially at all points on the boundary of this region. In fact, the solution of the equation is just the determination of the value of the function on part of the boundary, e.g. $t = t'$, a constant. Replacement of the differential equation by a difference equation proceeds as before. However, it is found that the convergence to a solution of the difference equation depends very much on the mesh adopted.

The most familiar of such equations is the diffusion equation, resulting from energy conservation in heat flow. At any point

$$\frac{\partial \varepsilon}{\partial t} + \operatorname{div} \boldsymbol{J} = 0$$

where ε and $\boldsymbol{J}$ are the energy density and heat flux vector respectively. When these quantities are proportional to the temperature (denoted by U for generality), and its gradient, respectively, we have an equation of the form $LU = 0$, with $L = \operatorname{div} \kappa \operatorname{grad} - \partial/\partial t$ where $\kappa \equiv k/\rho c_p$ is the thermal diffusivity of the medium. For the case of a one-dimensional rod of unit length, with κ constant and a specified initial temperature distribution $f(x)$, and its ends thereafter held at fixed temperatures U_0 and U_1, the problem is

$$\kappa \frac{\partial^2 U}{\partial x^2} - \frac{\partial U}{\partial t} = 0 \qquad U(x, 0) = f(x)$$

$$U(0, t) = U_0$$

$$U(1, t) = U_1. \tag{3.37}$$

The solution is just the value of U on the boundary in figure 3.8 at $t = T$.

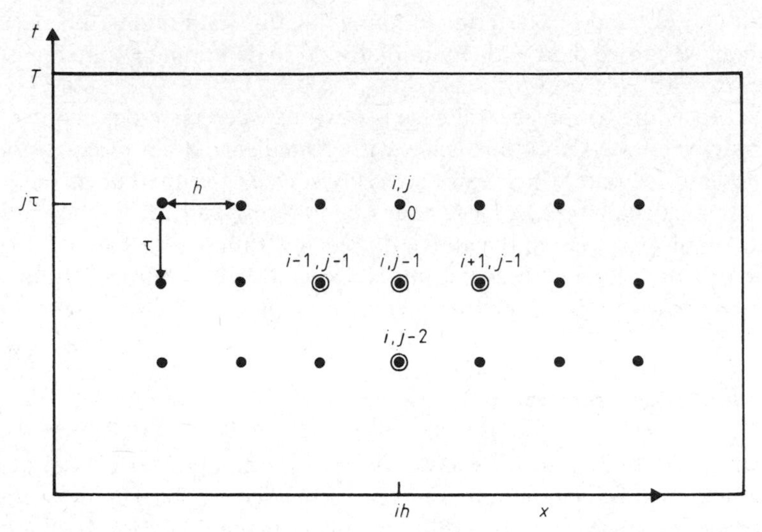

Figure 3.8 A possible set of nodal values occurring in an iterative solution, at the point 0, of equation (3.38).

Using mesh lengths h and τ for the two variables, we use (3.1) and (3.3) to replace (3.37) with a difference equation which has truncation errors $O(h^2)$ and $O(\tau^2)$ (see figure 3.8)

$$\frac{u_{i,j} - u_{i,j-2}}{2\tau} = \frac{\kappa}{h^2}(u_{i-1,j-1} - 2u_{i,j-1} + u_{i+1,j-1}). \tag{3.38}$$

If we have determined all the vlaues of u up to the time $(j-1)\tau$, we can, in principle, use (3.38) to find all the values at the next time step, $j\tau$, and hence find the solution at any finite time. An investigation of this method shows that no choice of h and τ guarantees a stable convergence to a solution; the values may start oscillating wildly from row to row. The simple replacement of $u_{i,j-1}$ on the right-hand side of (3.38) by its average in the t direction, i.e. $(u_{i,j-2} + u_{i,j})/2$, yields the *DuFort–Frankel* equation

$$u_{i,j} = \left(\frac{1-\alpha}{1+\alpha}\right)u_{i,j-2} + \left(\frac{\alpha}{1-\alpha}\right)(u_{i-1,j-1} + u_{i+1,j-1}) \tag{3.39}$$

where $\alpha = 2\kappa\tau/h^2$.

This explicit formula leads to a stable solution for any h and τ. However, it suffers from certain disadvantages; to get a small truncation error a small mesh will be required, since none of the four nearest neighbours of the point 0 in figure 3.8 are being used in equation (3.39) (compare with equation (3.18) for the elliptic case). Furthermore, since levels $(j-2)$ and $(j-1)$ enter into the formula, some different method will have to be used to establish the values $u_{i,j}$ when $j = 1$. These disadvantages can be overcome by using different methods for replacing the differential equation by a difference equation. However, in general, these lead to implicit formulae, i.e. a set of simultaneous equations must be solved at each time level; an important example is the Crank–Nicolson method for which the reader is referred to Ames (1977) and Lapidus and Pinder (1982).

Exercises

3.1 Verify equation (3.4), i.e. that the error committed by taking only the first term is of $O(h^2)$. If Φ is the solution of a differential equation and ϕ_1 and ϕ_2 are the solutions of the corresponding difference equations with mesh sizes h and h/j respectively, show that an improved approximation to Φ is $(j^2\phi_2 - \phi_1)/(j^2 - 1)$. This technique, known as Richardson extrapolation, can be used in any problem, e.g. numerical differentiation or integration, where the truncation error is $O(h^2)$.

3.2 If the eigenvectors $\boldsymbol{v}_i$ of the iteration matrix $\mathbf{J}$ in the general iteration formula $\boldsymbol{\phi} = \mathbf{J}\boldsymbol{\phi} + \boldsymbol{c}$ are used as a basis to expand the error vector $\boldsymbol{e}^{(0)}$, i.e. $\boldsymbol{e}^{(0)} = \Sigma\, c_i \boldsymbol{v}_i$, show that a necessary condition for the iteration to converge is that the maximum eigenvalue of $\mathbf{J}$ is less than unity. Provided this maximum eigenvalue $\rho(J)$ is not degenerate show that, at sufficiently large k, $\boldsymbol{e}^{(k)} \simeq \rho(J)\boldsymbol{e}^{(k-1)}$. See Chapter 5 of Smith (1978).

3.3 Improvement in the rate of convergence using SOR iteration may be achieved by (i) varying the factor ω on successive sweeps through the points from an initial value of 1 to its final asymptotic value $\omega_{\min}$, and (ii) carrying out the iterations in a leap-frog manner, i.e. first all the values ϕ_i, where i is odd, are relaxed and then, possibly using a different value for ω, all the points with even values of i are relaxed. Denoting the sequence of iterations on the odd and even points by $(k - \frac{1}{2})$ and k respectively, and denoting the value for ω in an SOR sweep through the points on such an iteration by $\omega^{(k)}$, the cyclic Chebyshev method consists of taking;

$$\omega^{(1/2)} = 1$$

$$\omega^{(1)} = 1/[1 - (\rho^2/2)]$$

$$\omega^{(k+1/2)} = 1/[1 - (\rho^2\omega^k/4)] \qquad k = 1, 3/2, 2\ldots$$

where ρ is the spectral radius of the Jacobi iteration matrix (Hockney 1970).

Develop a small program to compare this method with SOR for the following problem: $\nabla^2\Phi = 12x^2y^2(x^2 + y^2)$ in the unit square, with $\Phi(0, y) = \Phi(x, 0) = 0$, $\Phi(1, y) = y^4$, $\Phi(x, 1) = x^4$. Take a 4×4 mesh and use (3.17) to calculate ρ. Compare also the results with the known solution of the problem, namely $\Phi = x^4y^4$.

3.4 Find the capacitance per unit length normal to the page of the electrode configuration shown in figure 3.4 for the dimensions

AB = AP = CE = 4 cm, CD = 8 cm and DB = 12 cm, with air in the intervening space. (Answer: 20.3 pF.) Make a plot of the charge density on each plate as a function of position along it.

3.5 The diffusion equation for neutrons in a two-dimensional region of a reactor which is a square of side L is

$$\nabla^2\Phi - a^2\Phi + \sin\left(\frac{\pi x}{L}\right)\sin\left(\frac{\pi y}{L}\right) = 0.$$

Write down the corresponding finite difference equations for a square mesh of size h, and show that they are satisfied by

$$\phi_{ij} = \sin\left(\frac{i\pi h}{L}\right)\sin\left(\frac{j\pi h}{L}\right)\left[a^2 + \frac{8}{h^2}\sin^2\left(\frac{\pi h}{2L}\right)\right]^{-1}.$$

By comparison with the known solution

$$\Phi(x, y) = \left[\sin\left(\frac{\pi x}{L}\right)\sin\left(\frac{\pi y}{L}\right)\right]\left[a^2 + \left(\frac{2\pi^2}{L^2}\right)\right]^{-1}$$

show that an upper limit on the relative truncation error is $(\pi^2/12)(h/L)^2$.

3.6 Set up the difference equations, corresponding to Laplace's equation, for the three points on the unit mesh shown in figure 3.9, with the boundary conditions indicated there. Invert the 3×3 matrix to find the solutions. (Answers: 1727/1820, 73/104, 1563/1820.) Show that the spectral radius of the Jacobi iteration matrix is $1/\sqrt{14}$. Construct the Gauss–Seidel iteration matrix, $(\mathbf{I} - \mathbf{L})^{-1}\mathbf{U}$ and find its spectral radius. Determine the optimum parameter ω for over-relaxation and investigate the rate at which iterated values converge.

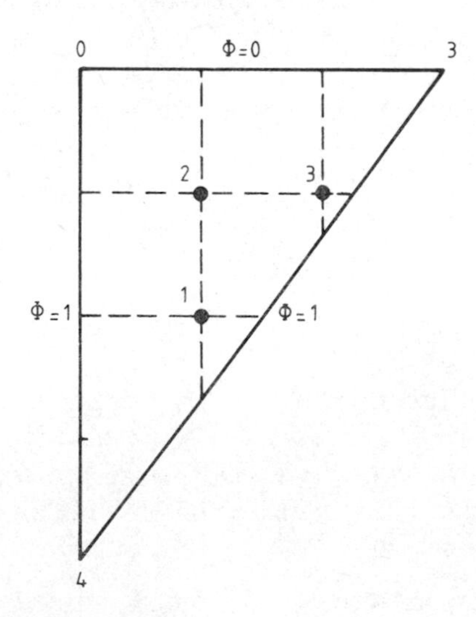

Figure 3.9 Mesh used in Exercise 3.6.

References

Ames W F 1977 *Numerical Methods for Partial Differential Equations* (New York: Academic)

Carre B A 1961 *Comput. J.* **4** 73–8

Fox L 1962 *Numerical Solution of Ordinary and Partial Differential Equations* (Oxford: Pergamon)

Hockney R W 1970 *Methods in Computational Physics* ed. B Alder *et al* vol. 9 (New York: Academic)

Lapidus L and Pinder G F 1982 *Numerical Solution of Partial Differential Equations in Science and Engineering* (New York: John Wiley)

Smith G D 1978 *Numerical Solution of Partial Differential Equations* (Oxford: Clarendon)

Tomitani T 1972 *Nucl. Instrum. Meth.* **100** 179–91

Urbanczyk K M and Waligorski M P R 1975 *Nucl. Instrum. Meth.* **124** 413–28

Chapter 4

The Matrix Eigenvalue Problem and its Application to Molecular Orbital Theory

4.1 Introduction

In this chapter we shall describe some simple computational methods of solving the matrix eigenvalue problem, namely finding the eigenvectors $\boldsymbol{x}_i$ and eigenvalues λ_i which simultaneously provide solutions to the equation

$$\mathbf{A}\boldsymbol{x} = \lambda \boldsymbol{x} \tag{4.1}$$

where $\mathbf{A}$ is a given matrix. We shall find that most of the physical applications of this equation involve an energy matrix $\mathbf{A}$ which is real and symmetric. We therefore restrict our analysis to such matrices.

It is necessary to distinguish several computational problems; 'complete' eigenvalue problems for which *all* of the $\boldsymbol{x}_i$ and λ_i are required and 'restricted' problems when it is only necessary to find a few particular solutions or when only eigenvalues, not eigenvectors, are required. The three methods we shall discuss in detail are each most appropriate to one of these types of computational problem. All are sufficiently simple to program directly on a microcomputer without the need for library subroutines. Finally, we shall briefly discuss some of the more sophisticated methods that the reader is likely to find implemented in library programs.

As examples of the application of these methods we have introduced two simple types of molecular orbital theory, relating to π-bonding in planar organic molecules and the bonding of simple diatomic molecules.

4.2 The Direct Method of Finding all Eigenvalues

As the heading indicates, this method proceeds to a direct solution of the characteristic equation $|\mathbf{A} - \lambda\mathbf{I}| = 0$, where $\mathbf{I}$ is the unit matrix and the bars indicate the determinant of the matrix $\mathbf{A} - \lambda\mathbf{I}$. The method is straightforward to program but is not usually discussed in books on numerical analysis because it is not fast for large matrices. Many problems in physics, however, do not involve the diagonalisation of large matrices, so we should take the method seriously.

It will be assumed that an attempt has already been made to express the matrix $\mathbf{A}$ in block diagonal form by permuting the rows and columns using a suitable permutation matrix $\mathbf{S}$ as follows (see (A.4))

$$\mathbf{SAS}^{\mathrm{T}} = \mathbf{A}' = \left[\begin{array}{c|c|c} \mathbf{A}_1 & \mathbf{0} & \mathbf{0} \\ \hline \mathbf{0} & \mathbf{A}_2 & \mathbf{0} \\ \hline \mathbf{0} & \mathbf{0} & \mathbf{A}_3 \end{array}\right]$$

Note that $\mathbf{S}^{\mathrm{T}} = \mathbf{S}^{-1}$. This transformation does not change the eigenvalues (as we shall show in §4.4) and only permutes the components of the eigenvector. Group representation theory (for example, see Leech and Newman (1969)) provides a systematic approach for reducing energy matrices to block diagonal form using the symmetry of the system. The mathematical advantage of such reductions is that they allow the factorisation of the secular determinant

$$|\mathbf{A} - \lambda\mathbf{I}| = |\mathbf{A}_1 - \lambda\mathbf{I}| \times |\mathbf{A}_2 - \lambda\mathbf{I}| \times |\mathbf{A}_3 - \lambda\mathbf{I}| \times \cdots$$

and hence break the original characteristic equation into several simpler equations.

The characteristic equation for an $n \times n$ square matrix $\mathbf{A}$ can be expressed as an nth degree polynomial equation

$$|\mathbf{A} - \lambda\mathbf{I}| = \lambda^n - a_1\lambda^{n-1} + a_2\lambda^{n-2} - \cdots - (-1)^n a_n = 0. \tag{4.2}$$

The first problem is to find an efficient method of relating the coefficients a_r to the elements of the matrix $\mathbf{A}$. We can then employ one of the standard methods of solving polynomial equations to find the eigenvalues (or roots) λ_i. In the case where $\mathbf{A}$ is symmetric and real, all the λ_i values can be shown to be real (see (A.4)). This greatly simplifies the problem of finding the roots.

The coefficients a_r may be shown to satisfy the following recurrence relations:

$$a_1 = \mathrm{Tr}(\mathbf{A})$$

$$a_2 = -\tfrac{1}{2}\,\mathrm{Tr}(\mathbf{A}^2 - a_1\mathbf{A})$$

and in general

$$a_r = (-1)^{r-1} r^{-1} \mathrm{Tr}(\mathbf{A}^r - a_1\mathbf{A}^{r-1} + a_2\mathbf{A}^{r-2} - \cdots - (-1)^{r-1} a_{r-1}\mathbf{A}). \tag{4.3}$$

Hence, the coefficients a_r can be found efficiently if we have a fast routine for calculating the powers of a real symmetric matrix $\mathbf{A}$ and sufficient storage to retain the traces of the $\mathbf{A}^r$ values as they are calculated. It is suggested that some thought be given to the construction of a special routine for this purpose.

Given the explicit form of equation (4.2), the next step is to find its roots. Many library programs exist which determine the roots of a polynomial equation. It is best to choose one that is specifically adapted for the calculation of *real* roots, otherwise time will be lost while it carries out unnecessary tests. If, on the other hand, readers cannot obtain a library program or wish to carry out some programming of their own, this problem provides an interesting exercise in writing a strongly interactive program for a microcomputer. We outline a possible procedure below.

Presuming that all the eigenvalues are required, the first step is to find the approximate location of all the roots of the characteristic function $f(\lambda)$ by plotting it over a sufficiently extensive range of values of λ. Generally speaking, this range will cover all positive and negative values of λ between those values for which the highest degree term in λ^n becomes dominant. It is useful to keep close control of this operation in order to ensure that the function is plotted in sufficient detail in all places and that only physically reasonable values of λ are scanned. A high-resolution graphical output is particularly helpful. Single and other odd multiplicity roots are easy to spot because $f(\lambda)$ changes sign. Double and other even multiplicity roots are more difficult as $f(\lambda)$ does not change sign. For this reason, it may be useful to plot the first differential $\mathrm{d}f(\lambda)/\mathrm{d}\lambda$ as well, although in practice symmetry considerations can be used to remove most degeneracies.

If the number of roots found in the above process is not yet equal to the degree of $f(\lambda)$, some of the roots must have a higher multiplicity than was apparent. Triple and higher multiplicity can be checked by examining the rth differentials of the characteristic function in the region of the root. If $\mathrm{d}^r f(\lambda)/\mathrm{d}\lambda^r$ has a root at the same place, then the root has multiplicity $(r+1)$.

Having determined approximate positions and multiplicities of all the roots of the characteristic equation, the second step is to refine their values to the desired accuracy. This can be done most simply by successive divisions of the interval in which the root is known to lie. A more popular method, which is only slightly more difficult to program, is based on Newton's rule: given an approximate root λ_0 of $f(\lambda)$, a better approximation is $\lambda_0 + \delta\lambda$, where

$$\delta\lambda = -f(\lambda_0)[(\mathrm{d}f/\mathrm{d}\lambda)_{\lambda_0}]^{-1}.$$

Reference to figure 4.1 will convince the reader that this formula iterates rapidly towards better values of the root (λ_1, λ_2 etc) if it is of multiplicity one. Roots of multiplicity $(r+1)$ can best be located by applying either of the iterative procedures described above for the function $\mathrm{d}^r f(\lambda)/\mathrm{d}\lambda^r$.

Interactive computing procedures are particularly useful in the approximate location of roots because the operator's expectations, based on a knowledge of the properties of the physical system, can be brought into play. In some problems, for example, only positive roots will be expected or there may be a known number of zero roots.

The reader must be warned that pathological polynomials exist for which the values of the roots are extremely sensitive to small changes in the co-

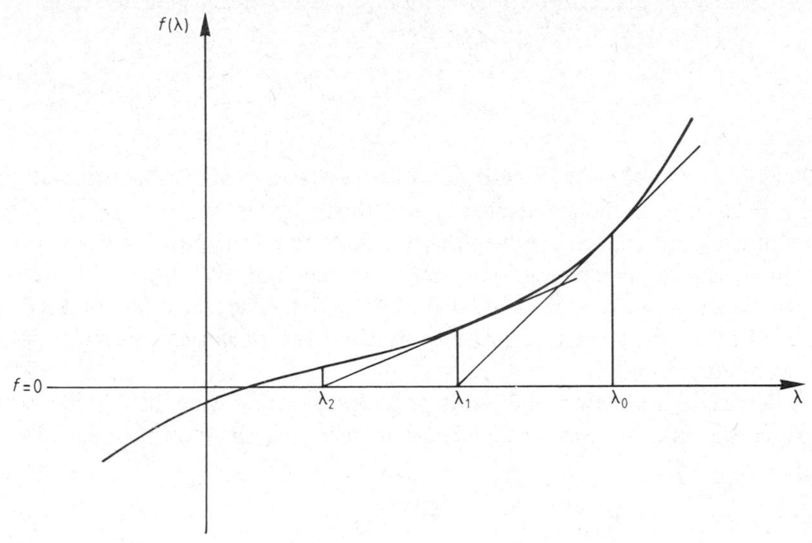

Figure 4.1 Two iterations of Newton's rule for finding the solutions of the equation $f(\lambda) = 0$ where the functional form is known.

efficients a_r. In particular, these are known to occur when the sequence of roots is equally spaced (for example, see Wilkinson (1965), p 418). Difficulties will arise in treating physical systems which have this property.

If it is necessary to find the eigenvectors as well as the eigenvalues, the most straightforward procedure is to solve the simultaneous equations

$$(\mathbf{A} - \lambda \mathbf{I})\boldsymbol{x} = 0$$

for the vectors $\boldsymbol{x} = \boldsymbol{x}_i$ corresponding to each $\lambda = \lambda_i$. In fact, only the ratios of vector components $\boldsymbol{x}_\alpha / \boldsymbol{x}_\beta$ can be determined from these equations, so it is necessary to begin by assuming that one of the components, $\boldsymbol{x}_\alpha$ (which you are convinced will not be zero in the solution), takes the value one. Then one of the many standard routines for solving simultaneous equations can be used.

4.3 The Power Method of Finding a Few Eigenvectors and Eigenvalues

This method has been described in numerous treatises on numerical methods (Acton 1970, Nash 1979) and yet its simplicity and suitability for interactive use of the computer merit it a place in this book. It is necessary to start with an informed guess the form of the eigenvector (denoted $\boldsymbol{x}_0$) which corresponds to the eigenvalue with the largest modulus. In physical applications it is often possible to use the properties of the system under discussion to make this guess reasonable.

The iterative procedure for solving equation (4.1) may be written

$$\begin{aligned} \boldsymbol{y}_{\alpha+1} &= \mathbf{A}\boldsymbol{x}_{\alpha} \\ \boldsymbol{x}_{\alpha+1} &= \frac{\boldsymbol{y}_{\alpha+1}}{|\boldsymbol{y}_{\alpha+1}|} \end{aligned} \tag{4.4}$$

where $|\boldsymbol{y}_{\alpha}|$ represents the modulus of the vector $\boldsymbol{y}_{\alpha}$. It is continued until $\boldsymbol{x}_{\alpha+1} = \boldsymbol{x}_{\alpha}$ to a sufficient accuracy for the problem in hand. Rather than dividing by $|\boldsymbol{y}_{\alpha}|$ at each iteration, it is possible to adopt any reasonable renormalisation which is quick to compute, such as dividing the elements of $\boldsymbol{y}_{\alpha}$ by their maximum absolute value. When convergence is obtained, the eigenvalue is given most accurately by the ratio of largest (corresponding) elements in $\boldsymbol{x}_{\alpha}$ and $\boldsymbol{y}_{\alpha}$.

In order to understand how this procedure works, we consider the expansion of the first 'guessed' eigenvector in terms of the complete set of exact eigenvectors $\boldsymbol{x}_m$:

$$\boldsymbol{x}_0 = \sum_m \boldsymbol{x}_m b_m$$

where the b_m values are expansion coefficients. Multiplying on the left by $\mathbf{A}^{\alpha}$, where α is an integral exponent, we obtain

$$\mathbf{A}^{\alpha}\boldsymbol{x}_0 = \sum_m \boldsymbol{x}_m \lambda_m^{\alpha} b_m$$

showing that the vector $\mathbf{A}^{\alpha}\boldsymbol{x}_0$ has an enhanced contribution to the $\boldsymbol{x}_m$ component corresponding to the largest value $|\lambda_m|_{\max}$ of the $|\lambda_m|$. A sufficiently large value of α will thus produce a vector $\mathbf{A}^{\alpha}\boldsymbol{x}_0$ closely proportional to the eigenvector for $|\lambda_m|_{\max}$.

The rate of convergence of the iterative process can often be improved by the simple device of shifting all the eigenvalues by the same amount, Δ. This is achieved by the transformation

$$\mathbf{A} \rightarrow \mathbf{A} - \Delta\mathbf{I} = \mathbf{A}'$$

where $\mathbf{I}$ is the unit matrix. $\mathbf{A}'$ has eigenvalues $(\lambda_i - \Delta)$. Having found the largest eigenvalue $\lambda_{\max}$ it is easy to find the eigenvalue at the opposite end of the ordered sequence by putting $\Delta = \lambda_{\max}$ and solving for $\mathbf{A}'$.

Another trick is to iterate using $\mathbf{A}^{-1}$ to find the smallest eigenvalue of $\mathbf{A}$. This does, however, add significantly to the computation time. Combining these ideas, it is possible to obtain the eigenvalue nearest to an arbitrary value q by iterating with $(\mathbf{A} - q\mathbf{I})^{-1}$.

It will be apparent from the above discussion that the power method is most appropriate if we wish to determine relatively few eigenvalues (and their corresponding eigenvectors). Nevertheless, it can be applied to determine the complete set of eigenvalues and eigenvectors if a systematic method is used to subtract that part of the matrix elements which contributes to the eigenvalues that have already been determined. This may be done by replacing the matrix $\mathbf{A}$ with $\mathbf{A} - \lambda_i \boldsymbol{x}_i \boldsymbol{x}_i^{\mathrm{T}}$ after each eigenvector $\boldsymbol{x}_i$ and eigenvalue λ_i has been determined. (T indicates transposition and we are assuming that $\boldsymbol{x}_i$ is normalised so that $\boldsymbol{x}_i^{\mathrm{T}} \boldsymbol{x}_i = 1$.) This procedure is easy to program, but must be used with care because of the ease of accumulating errors when the largest contribution to the matrix elements is subtracted at each stage.

4.4 Diagonalisation Procedures

These are based on finding a systematic sequence of transformations which will bring the matrix $\mathbf{A}$ into diagonal form. There are several very efficient methods of this type that the reader is most likely to encounter in library programs. In any orthogonal transformation of the matrix $\mathbf{A}$,

$$\mathbf{A}' = \mathbf{U}^{\mathrm{T}} \mathbf{A} \mathbf{U} \tag{4.5}$$

where $\mathbf{U}^{\mathrm{T}}\mathbf{U} = \mathbf{I}$, the matrix $\mathbf{A}'$ has the same eigenvalues as $\mathbf{A}$, but its eigenvectors are transformed by the matrix $\mathbf{U}$ (see equation (A.24) in the Appendix).

The Jacobi method (Wilkinson 1965, Acton 1970, Nash 1979) seeks to eliminate all the off-diagonal elements of $\mathbf{A}$ by the successive diagonalisation of 2×2 submatrices. The transformation matrix for eliminating the (i, j)th element takes the form

$$\mathbf{U}(i, j; \beta) = \begin{array}{c} \\ \\ \\ \\ i \\ \\ \\ j \\ \\ \\ \end{array} \left[\begin{array}{ccccccccc} 1 & & & & & & & & \\ & \ddots & & & & & & & \\ & & 1 & & & & & & \\ & & & \cos\beta & & & \sin\beta & & \\ & & & & 1 & & & & \\ & & & & & 1 & & & \\ & & & -\sin\beta & & & \cos\beta & & \\ & & & & & & & 1 & \\ & & & & & & & & 1 \end{array} \right]$$

(columns i and j as marked above the matrix)

where the angle β is determined by the relation

$$\tan 2\beta = \frac{2A_{ij}}{A_{ii} - A_{jj}}. \tag{4.6}$$

This process is iterative because all other off-diagonal elements in rows and columns i and j are changed, in addition to A_{ij}. The total orthogonal transformation matrix, corresponding to a product of the matrices $\mathbf{U}(i, j; \beta)$, must be accumulated if it is wished to reconstruct the eigenvectors of $\mathbf{A}$.

In early applications of the Jacobi method using hand calculators it was important to speed the process by removing the largest off-diagonal element in each step. This is not necessarily the quickest method for a large computer because of the time spent in determining which element is the largest, so that computer programs usually scan sequentially through the off-diagonal matrix elements, regardless of size. The reader might be interested to compare the relative speed of these procedures using an eight-bit microcomputer with (*a*) sequential scanning, (*b*) largest element selection by computer and (*c*) largest element selection by the operator from a screen display.

Givens has modified the Jacobi method (see Wilkinson (1965)), using the criterion

$$\tan \beta = A_{i-k,j}/A_{i-k,i} \tag{4.7}$$

to eliminate the matrix element at $(i - k, j)$ rather than at (i, j). This has the advantage that it can be applied systematically, row-by-row, until all elements are zero except those on the main diagonal and on either side of it. The disadvantage is that it only produces a tridiagonal rather than a diagonal matrix. This may, however, be used to form the characteristic function using a simple recurrence relation (for details see Acton 1970, Chapter 13).

Householder (Wilkinson 1965, Acton 1970, Nash 1979) has introduced a speeded up form of Givens' method in which a transformation matrix is constructed in such a way as to eliminate all possible elements of a given row at the same time. The final result of the transformation process is again a tridiagonal matrix. Householder's method clearly requires more programming but has been shown to be significantly faster in operation than the Givens' method. It is frequently the method preferred for library programs.

4.5 Molecular Orbital Theory

Molecular orbitals are single electron states of molecules: they are usually constructed from atomic orbitals. The outer atomic, ns^2p^6, shells may be filled, in which case they contain eight electrons corresponding to the four orbital and two spin degrees of freedom, or they may be partially filled. Atoms with outer filled shells cannot stabilise by sharing electrons with other atoms and therefore do not form molecules. In the simplest form of molecular orbital theory, one-electron orbitals on a system consisting of a group of atoms are constructed from linear combinations of the orbitals in the inner filled and outer partially filled atomic shells. Even such a severely restricted basis can provide a reasonable approximation to the low-lying energy levels of the system.

Supposing the unfilled shell atomic orbitals are denoted by ϕ_i, we may write the states of the complete system as

$$\psi = \sum_i a_i \phi_i. \tag{4.8}$$

This method of construction of *molecular orbitals* by taking *linear combinations of atomic orbitals* is often referred to in the literature as the MO–LCAO method. The coefficients a_i in equation (4.8) may be determined by the *variational principle* or by solving the Schrödinger equation over the restricted basis set. We write this (for the one-electron states) in the form

$$H\psi = E\psi \tag{4.9}$$

where H is the energy operator and the eigenvalues E correspond to the eigenstates ψ in the solutions of the equation. Substituting the trial solution (4.8) into (4.9) we obtain

$$\sum_i a_i H\phi_i = E \sum_i a_i \phi_i. \tag{4.10}$$

Multiplying on the left by ϕ_k^* and integrating over electron coordinates we have

$$\sum_i a_i \int \phi_k^* H\phi_i \, \mathrm{d}\tau = E \sum_i a_i \int \phi_k^* \phi_i \, \mathrm{d}\tau \tag{4.11}$$

or

$$\sum_i a_i H_{ki} = E \sum_i a_i S_{ki} \tag{4.12}$$

where H_{ki} and S_{ki} denote matrix elements of the Hamiltonian and the overlap matrix respectively. In matrix form (4.12) becomes

$$\mathbf{H}\boldsymbol{a} = E\mathbf{S}\boldsymbol{a}. \tag{4.13}$$

Exercise

4.1 Prove that the variational principle also leads to equation (4.13) for the coefficients a_i.

The problem of finding solutions to equation (4.13) is known as the *generalised eigenvalue problem*. We can normally choose our orbitals to be real functions so that $\mathbf{H}$ and $\mathbf{S}$ are symmetric. The trivial transformation to $\mathbf{S}^{-1}\mathbf{H}a = Ea$ gives an asymmetric matrix eigenvalue problem which leads to a considerable complication of the numerical procedures. We therefore seek a method which converts equation (4.13) into an eigenvalue problem for symmetric matrices.

The normalisation integral $\langle\psi|\psi\rangle = \boldsymbol{a}^{\mathrm{T}}\mathbf{S}\boldsymbol{a}$ is greater than 0, so that $\mathbf{S}$ is positive definite, i.e. all its eigenvalues are positive. Writing them as D_{ii}^2, $\mathbf{S}$ may be expressed as

$$\mathbf{S} = \mathbf{Z}\mathbf{D}^2\mathbf{Z}^{\mathrm{T}}$$

where $\mathbf{Z}$ is an orthogonal transformation so that $\mathbf{Z}\mathbf{Z}^{\mathrm{T}} = \mathbf{I}$. Note that $\mathbf{Z}$ is constructed from the eigenvectors of $\mathbf{S}$. Given $\mathbf{Z}$ and $\mathbf{D}^2$, therefore, we may write

$$\mathbf{S}^{\pm 1/2} = \mathbf{Z}\mathbf{D}^{\pm 1}\mathbf{Z}^{\mathrm{T}} \tag{4.14}$$

where we take $D_{ii} > 0$ and transform (4.13) to the standard eigenvalue problem:

$$(\mathbf{S}^{-1/2}\mathbf{H}\mathbf{S}^{-1/2})(\mathbf{S}^{1/2}\boldsymbol{a}) = E(\mathbf{S}^{1/2}\boldsymbol{a}). \tag{4.15}$$

In this way equation (4.13) is reduced to two ordinary eigenvalue problems, first for $\mathbf{S}$ and then for $\mathbf{S}^{-1/2}\mathbf{H}\mathbf{S}^{-1/2}$. Hence, a solution may be found by successive applications of a standard eigenvalue program. It should be remarked that this is not the most efficient numerical way to solve this problem (see Wilkinson 1965), but it is nevertheless the most convenient in the present context.

4.6 Planar Molecules: Hückel Theory

A particularly simple version of molecular orbital theory is the Hückel theory, as applied to π-bonding in planar organic molecules (Yates 1978). The physical basis of this theory is that the bonding between pπ-orbitals orientated perpendicular to the molecular plane is relatively weak, so that the energy gaps between the occupied and unoccupied states are small. Hence, the lowest energy electronic excitations of the system will be between occupied and unoccupied states constructed from atomic pπ orbitals. In the case of the planar molecule butadiene, which is normally represented by $CH_2{=}CH{-}CH{=}CH_2$, π-bonding produces the second bond in the double bonds. We may use Hückel theory to determine whether this conventional assignment of double bonds really represents the electronic distribution.

In planar organic molecules, such as butadiene, a single pπ-electron may be assigned to each carbon atom, in addition to the underlying single bonded structure or 'skeleton', e.g. $CH_2{-}CH{-}CH{-}CH_2$. The Hückel theory associates each such π-electron with a self-energy $H_{ii} = \alpha$ and an interaction energy between nearest neighbours, $H_{ij} = \beta$ (less than zero). Hence, in the case of butadiene, the energy matrix takes the tridiagonal form

$$\mathbf{H} = \begin{bmatrix} \alpha & \beta & 0 & 0 \\ \beta & \alpha & \beta & 0 \\ 0 & \beta & \alpha & \beta \\ 0 & 0 & \beta & \alpha \end{bmatrix} \tag{4.16}$$

It is also assumed that the π-electron orbitals are orthogonal. We define $x = (\alpha - E)/\beta$, so the secular equation may be written $x^4 - 3x^2 + 1 = 0$ which (solving for x^2) gives $x = \pm 1.62$ and ± 0.62. The four electrons are therefore spin paired in the two lowest lying energy states. These correspond to the eigenvectors

$$\begin{aligned} \psi_A &= 0.37\phi_1 + 0.60\phi_2 + 0.60\phi_3 + 0.37\phi_4 \quad (E_A = \alpha + 1.62\beta) \\ \psi_B &= 0.60\phi_1 + 0.37\phi_2 - 0.37\phi_3 - 0.60\phi_4 \quad (E_B = \alpha + 0.62\beta). \end{aligned} \tag{4.17}$$

In these equations ϕ_i is the π-electron function on the ith carbon atom.

Neglecting electron–electron interactions, the total energy is therefore given by $2(E_A + E_B) = 4\alpha + 4.48\beta$. This may be compared with the energy $4\alpha + 4\beta$ for the system corresponding to the classical double-bonded picture of butadiene, for which

$$\begin{aligned} \psi_A &= \tfrac{1}{2}(\phi_1 + \phi_2 + \phi_3 + \phi_4) \\ \psi_B &= \tfrac{1}{2}(\phi_1 + \phi_2 - \phi_3 - \phi_4) \end{aligned} \tag{4.18}$$

so that

$$\psi_A^2 + \psi_B^2 = \tfrac{1}{2}(\phi_1^2 + \phi_2^2 + \phi_3^2 + \phi_4^2 + 2\phi_1\phi_2 + 2\phi_3\phi_4).$$

This may be compared with the expression

$$\psi_A^2 + \psi_B^2 = \tfrac{1}{2}(\phi_1^2 + \phi_2^2 + \phi_3^2 + \phi_4^2) + 0.89(\phi_1\phi_2 + \phi_3\phi_4) + 0.45(\phi_2\phi_3 + \phi_1\phi_4)$$

obtained from equation (4.17) above, showing that the qualitative features of double bonding of the atom pairs 1–2 and 3–4 are preserved in the solution of the Hückel model.

The symmetry properties of a molecule such as butadiene may be employed to simplify the eigenvalue problem, although this is hardly necessary for such a small matrix! The idea is to separate the odd and even solutions under the inversion operation. This may be achieved by the orthogonal transformation

$$\frac{1}{\sqrt{2}} \begin{bmatrix} 1 & 0 & 0 & 1 \\ 0 & 1 & 1 & 0 \\ 0 & -1 & 1 & 0 \\ -1 & 0 & 0 & 1 \end{bmatrix} \begin{bmatrix} \alpha & \beta & 0 & 0 \\ \beta & \alpha & \beta & 0 \\ 0 & \beta & \alpha & \beta \\ 0 & 0 & \beta & \alpha \end{bmatrix}$$

$$\times \frac{1}{\sqrt{2}} \begin{bmatrix} 1 & 0 & 0 & -1 \\ 0 & 1 & -1 & 0 \\ 0 & 1 & 1 & 0 \\ 1 & 0 & 0 & 1 \end{bmatrix} = \begin{bmatrix} \alpha & \beta & 0 & 0 \\ \beta & \alpha + \beta & 0 & 0 \\ 0 & 0 & \alpha - \beta & \beta \\ 0 & 0 & \beta & \alpha \end{bmatrix}$$

which block diagonalises the energy matrix corresponding to the factorisation of the secular equation $(x^2+x-1)(x^2-x-1)=0$. In such simple cases, therefore, it may be sufficient just to look for an expected factorisation of the secular equation.

A more interesting relationship with the classical picture of bonds and double bonds is provided by the cyclic structure of six carbon atoms in the hexagonal structure of benzene. In this case the energy matrix takes the form

$$\mathbf{H} = \begin{bmatrix} \alpha & \beta & 0 & 0 & 0 & \beta \\ \beta & \alpha & \beta & 0 & 0 & 0 \\ 0 & \beta & \alpha & \beta & 0 & 0 \\ 0 & 0 & \beta & \alpha & \beta & 0 \\ 0 & 0 & 0 & \beta & \alpha & \beta \\ \beta & 0 & 0 & 0 & \beta & \alpha \end{bmatrix}$$

and can be totally diagonalised by symmetry considerations. As an exercise, the reader is invited to determine the secular equation and then to seek the factorisation that is to be expected as a result of symmetry. Is the alternate double-bonding picture shown in the diagram correct?

Exercise

4.2 The naphthalene molecule has the double ring structure

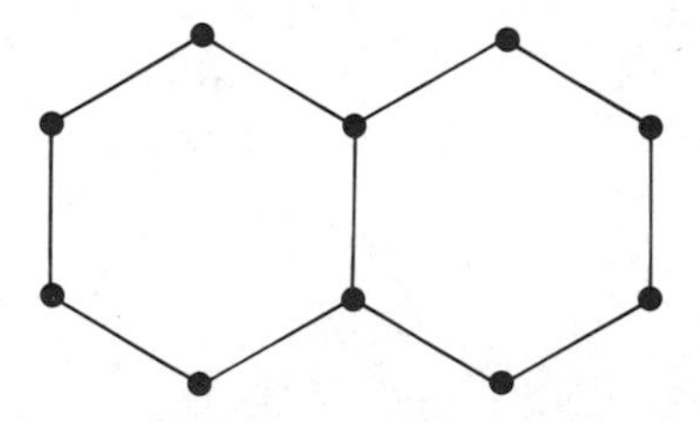

with 10 carbon atoms at the points indicated in the diagram. Write down the secular equation for this system and check for factorisations arising from symmetry. Show that only quadratic equations have to be solved. Using the eigenfunctions, suggest an approximate arrangement of double bonds which describes the electron distribution.

Further details of this topic may be found in Yates (1978).

4.7 Molecular Orbital Theory for Diatomic Molecules

The aim of the computation in this case is rather different to that discussed above for π-bonding in large molecules. In particular, we are interested in getting accurate numerical predictions of physical quantities such as the interatomic spacing, the bonding energy and the electron ionisation energy. For sufficiently light atoms (with a small number of electrons), it is possible to include all of the occupied atomic orbitals in the molecular orbital expression. In larger atoms it is usual to regard electrons in the core orbitals as fixed.

The simplest possible example is the hydrogen molecule in which each atom has a single 1s electron. Within this restricted basis approximate eigenstates are the antisymmetric (A) and symmetric (S) wavefunctions

$$\psi_A = N_A(\phi_1 - \phi_2)$$
$$\psi_S = N_S(\phi_1 + \phi_2)$$

where ϕ_1, ϕ_2 represent the two (non-orthogonal) atomic orbitals and N_A and N_S are constants chosen to normalise ψ_A and ψ_S. The term ψ_S is usually referred to as the *bonding orbital* and ψ_A as the *antibonding orbital.* This is because the electron density corresponding to ψ_S

$$|\psi_S|^2 = N_S^2(|\phi_1|^2 + |\phi_2|^2 + \phi_1^*\phi_2 + \phi_2^*\phi_1)$$

has a negative exchange charge distribution between the atomic centres, which tends to bond the two atoms, giving lower one-electron energies for ψ_S than ψ_A. Hence, in the present case, both available electrons will go into the bonding orbital, one with spin up and the other with spin down.

PROJECT 4A: LiH—A SIMPLE SELF-CONSISTENT FIELD PROJECT

The project is based on an article by Rioux and Harriss (1980) which is appended to this section. We have also provided (in table 4.1) the relevant matrix elements calculated by Karo and Olsen (1959). A related article by Karo (1959) also gives useful information. The main feature of the project is that the energy matrix has to be calculated 'self-consistently'—that is, it depends to some extent on its own eigenvectors. The physical interpretation of self-consistency is that the state of a given electron depends on the charge distribution and hence the states of the other electrons. The mathematical consequences are apparent in equation (8) of Rioux and Harriss, where an energy matrix element is shown to be explicitly dependent on the coefficients c in the wavefunction expansion. Such self-consistency can, of course, only be achieved by an iterative procedure.

The basis set used for this system consists of the atomic (1s) hydrogen orbital denoted ϕ_h and the outer 2s and 2p orbitals of Li, denoted ϕ_s and ϕ_p.

Table 4.1 One- and two-centre integrals for LiH at various atomic separations, R, taken from Karo and Olsen (1959).

One-centre integrals							
One-electron integrals				Two-electron integrals			
$[s/-(\nabla^2/2)/s]$	3.61037	$[s/(1/r)/s]$	2.68449	$ssss$	1.64978	$pSps$	−0.00492
$[S/-(\nabla^2/2)/s]$	0.69533	$[S/(1/r)/s]$	0.27275	$Ssss$	0.12216	$pSpS$	0.04403
$[S/-(\nabla^2/2)/S]$	0.20834	$[S/(1/r)/S]$	0.34550	$SsSs$	0.01412	$ppss$	0.26411
$[p/-(\nabla^2/2)/p]$	0.14100	$[p/(1/r)/p]$	0.26499	$SSss$	0.32298	$ppSs$	0.00101
$[h/-(\nabla^2/2)/h]$	0.50000	$[h/(1/r')/h]$	1.00000	$SSSs$	0.00278	$ppSS$	0.20881
				$SSSS$	0.23448	$pppp$	0.20403
				$psps$	0.00294	$hhhh$	0.62500

Two-centre one-electron integrals									
R (atomic units):	$R = 2.0$	$R = 2.6$	$R = 3.0$	$R = 3.5$	$R = 4.0$	$R = 5.0$	$R = 6.0$	$R = 7.0$	$R = 8.0$
(h/s)	0.24765	0.14903	0.10445	0.06613	0.04144	0.01595	0.00604	0.00227	0.00085
(h/S)	−0.53434	−0.50141	−0.46932	−0.42210	−0.37115	−0.27160	−0.18769	−0.12411	−0.07932
(h/p)	0.44962	0.49562	0.50598	0.50073	0.48017	0.41181	0.33017	0.25213	0.18579
$[h/-(\nabla^2/2)/s]$	0.03493	−0.00084	−0.00798	−0.00951	−0.00805	−0.00422	−0.00187	−0.00077	−0.00031
$[h/-(\nabla^2/2)/S]$	−0.05182	−0.05431	−0.04797	−0.03526	−0.02582	−0.00923	−0.00431	−0.00212	0.00281
$[h/-(\nabla^2/2)/p]$	0.11689	0.10922	0.09785	0.07853	0.06416	0.03641	0.02590	0.00864	0.00339
$[h/(1/r)/s]$	0.33138	0.18998	0.12998	0.08037	0.04947	0.01857	0.00692	0.00257	0.00095
$[h/(1/r)/S]$	−0.18650	−0.16662	−0.14885	−0.12533	−0.10276	−0.06523	−0.03927	−0.02280	−0.01290
$[h/(1/r)/p]$	0.17053	0.16696	0.15791	0.14253	0.12517	0.09110	0.06290	0.04191	0.02725
$[h/(1/r)/h]$	0.47253	0.37698	0.33003	0.28454	0.24958	0.19995	0.16666	0.14286	0.12500

Table 4.1 *(contd)*

$[s/(1/r')/s]$	0.49987	0.38453	0.33332	0.28571	0.25000	0.20000	0.16667	0.14286	0.12500
$[S/(1/r')/s]$	0.00216	0.00037	0.00010	0.00001	0.00000	0.00000	0.00000	0.00000	0.00000
$[S/(1/r')/S]$	0.29593	0.27859	0.26443	0.24549	0.22658	0.19216	0.16409	0.14203	0.12473
$[p/(1/r')/s]$	0.02245	0.01457	0.01120	0.00832	0.00639	0.00410	0.00285	0.00209	0.00160
$[p/(1/r')/S]$	−0.09275	−0.10004	−0.09978	−0.09563	−0.08891	−0.07285	−0.05771	−0.04540	−0.03596
$[p/(1/r')/p]$	0.28854	0.28414	0.27729	0.26541	0.25109	0.21970	0.18951	0.16333	0.14174
$[h/(1/r')/s]$	0.15876	0.07368	0.04424	0.02355	0.01267	0.00376	0.00115	0.00036	0.00012
$[h/(1/r')/S]$	−0.31899	−0.30501	−0.28263	−0.24631	−0.21139	−0.14503	−0.09815	−0.05994	−0.03685
$[h/(1/r')/p]$	0.34170	0.35704	0.35084	0.32889	0.30424	0.24231	0.19098	0.13470	0.09628

Two-centre two-electron integrals

R (atomic units):	$R = 2.0$	$R = 2.6$	$R = 3.0$	$R = 3.5$	$R = 4.0$	$R = 5.0$	$R = 6.0$	$R = 7.0$	$R = 8.0$
$h\,s\,s\,s$	0.26846	0.15526	0.10665	0.06619	0.04085	0.01539	0.00575	0.00214	0.00079
$h\,s\,S\,s$	0.01362	0.00762	0.00515	0.00315	0.00192	0.00071	0.00026	0.00010	0.00004
$h\,s\,S\,S$	0.07699	0.04611	0.03222	0.02033	0.01271	0.00487	0.00184	0.00069	0.00026
$h\,s\,p\,s$	0.00283	0.00174	0.00122	0.00077	0.00048	0.00018	0.00007	0.00003	0.00001
$h\,s\,p\,S$	−0.00526	−0.00360	−0.00266	−0.00177	−0.00114	−0.00046	−0.00018	−0.00007	−0.00003
$h\,s\,p\,p$	0.06522	0.03923	0.02747	0.01737	0.01087	0.00418	0.00158	0.00059	0.00022
$h\,s\,h\,s$	0.05453	0.01876	0.00897	0.00350	0.00135	0.00019	0.00003	0.00000	0.00000
$h\,S\,s\,s$	−0.19471	−0.17113	−0.15187	−0.12717	−0.10387	−0.06564	−0.03942	−0.02286	−0.01292
$h\,S\,S\,s$	0.00019	0.00002	−0.00001	−0.00003	−0.00002	−0.00001	−0.00001	0.00000	0.00000
$h\,S\,S\,S$	−0.13961	−0.12728	−0.11625	−0.10088	−0.08520	−0.05693	−0.03564	−0.02129	−0.01229
$h\,S\,p\,s$	−0.00416	−0.00387	−0.00347	−0.00288	−0.00230	−0.00136	−0.00075	−0.00040	−0.00021
$h\,S\,p\,S$	0.02675	0.02901	0.02874	0.02690	0.02398	0.01704	0.01088	0.00645	0.00362
$h\,S\,p\,p$	−0.12770	−0.11941	−0.11093	−0.09827	−0.08462	−0.05845	−0.03753	−0.02281	−0.01331
$h\,S\,h\,s$	−0.05384	−0.02889	−0.01810	−0.00965	−0.00496	−0.00121	−0.00027	−0.00006	−0.00001
$h\,S\,h\,S$	0.10869	0.09722	0.08523	0.06846	0.05229	0.02706	0.01244	0.00524	0.00207

Table 4.1 (*contd*)

	Two-centre two-electron integrals								
R (atomic units):	$R = 2.0$	$R = 2.6$	$R = 3.0$	$R = 3.5$	$R = 4.0$	$R = 5.0$	$R = 6.0$	$R = 7.0$	$R = 8.0$
h p s s	0.16985	0.16657	0.15765	0.14237	0.12507	0.09107	0.06289	0.04190	0.02725
h p S s	0.00102	0.00065	0.00046	0.00030	0.00019	0.00007	0.00003	0.00001	0.00000
h p S S	0.11551	0.12243	0.12120	0.11490	0.10514	0.08137	0.05851	0.04003	0.02648
h p p s	0.00625	0.00543	0.00479	0.00398	0.00323	0.00203	0.00122	0.00072	0.00041
h p p S	−0.03694	−0.03905	−0.03902	−0.03747	−0.03463	−0.02688	−0.01897	−0.01250	−0.00786
h p p p	0.11079	0.12004	0.12061	0.11643	0.10837	0.08629	0.06333	0.04387	0.02918
h p h s	0.04882	0.02906	0.01934	0.01107	0.00609	0.00169	0.00044	0.00011	0.00003
h p h S	−0.10273	−0.10452	−0.09835	−0.08565	−0.07061	−0.04228	−0.02241	−0.01088	−0.00494
h p h p	0.10411	0.11657	0.11660	0.10938	0.09700	0.06706	0.04100	0.02293	0.01202
h h s s	0.46641	0.37517	0.32923	0.28417	0.24920	0.19983	0.16717	0.14280	0.12479
h h S s	0.00469	0.00177	0.00089	0.00037	0.00015	0.00002	0.00000	0.00000	0.00000
h h S S	0.27832	0.26305	0.25123	0.23536	0.21939	0.18878	0.16317	0.14143	0.12434
h h p s	0.01453	0.01158	0.00970	0.00770	0.00614	0.00406	0.00285	0.00209	0.00160
h h p S	−0.07217	−0.08058	−0.08252	−0.08165	−0.07827	−0.06721	−0.05518	−0.04420	−0.03541
h h p p	0.26186	0.25868	0.25398	0.24547	0.23510	0.20993	0.18464	0.16052	0.14020
h h h s	0.13009	0.06552	0.04084	0.02244	0.01229	0.00372	0.00115	0.00036	0.00012
h h h S	−0.24751	−0.23450	−0.21847	−0.19383	−0.16769	−0.11762	−0.07821	−0.04957	−0.03056
h h h p	0.24833	0.26349	0.26249	0.25224	0.23587	0.19249	0.14839	0.10906	0.07807

In the simplified calculation described by Rioux and Harriss, the 1s Li orbital in the molecule is assumed to be unperturbed from its atomic form. The molecular orbital eigenfunctions therefore take the form shown in equation (3) of Rioux and Harriss, and the energy matrix is 3×3.

Note that the lithium atom has only three electrons, two of which are in 1s orbitals, so that only one electron is contributed to the molecular orbitals by each atom and hence both electrons go into the lowest energy state (with opposite spins). The matrix elements tabulated by Karo and Olsen are given for several interatomic spacings, so that an equilibrium distance between the Li and H atoms may, in principle, be determined and compared with the experimental value. In order to calculate the equilibrium distance it is necessary to determine an explicit expression for the total energy of the system corresponding to the self-consistent solutions. This is an interesting exercise in itself and takes the project a step further than was suggested by Rioux and Harriss. In order to reduce the time required to input data and reach a self-consistent solution, this project may best be treated as a group project, with each student obtaining a result for just one atomic spacing. Alternatively the data may be supplied to students in a form that can be read directly into the computer. (It should be noted, however, that with the restricted basis set suggested by Rioux and Harriss an energy minimum is not obtained in practice.)

As the two available electrons occupy the molecular orbital of lowest energy, self-consistency will be achieved by using the values of c_p, c_s and c_h corresponding to this eigenvalue. Convergence may be difficult to obtain (especially for the larger spacings) when a simple iterative approach is used. This problem may be overcome by using a weighted average of the *two* last solutions in the Hamiltonian matrix for the next iteration.

Note that (see table 4.1) Karo and Olsen write S instead of the symbol s used by Rioux and Harriss and s instead of s′. Several of the one-centre two-electron matrix elements in the expressions given by Rioux and Harriss are identically zero, e.g. (SS|Sp), and these will, of course, not be found in the Karo and Olsen tabulation.

References

The book by Leech and Newman (1969) concerns the possible use of symmetry properties to simplify numerical calculations. Wilkinson (1965), Acton (1970) and Nash (1979) are relevant to the calculation of eigenvalues and eigenvectors, Wilkinson's book being the standard work on this subject. Yates (1978) concerns planar organic molecules. The papers by Rioux and Harriss (1980), Karo and Olsen (1959) and Karo (1959) are related to the LiH project.

Acton F S 1970 *Numerical Methods that Work* (New York: Harper and Row)

Karo A M 1959 *J. Chem. Phys.* **30** 1241

Karo A M and Olsen A R 1959 *J. Chem. Phys.* **30** 1232

Leech J W and Newman D J 1969 *How to Use Groups* (London: Chapman and Hall)

Nash J C 1979 *Compact Numerical Methods for Computers* (Bristol: Adam Hilger)

Rioux F and Harriss D K 1980 *Am. J. Phys.* **48** 439

Wilkinson J H 1965 *The Algebraic Eigenvalue Problem* (Oxford: Clarendon)

Yates K 1978 *Hückel Molecular Orbital Theory* (New York: Academic)

Appendix 4A: Self-consistent-field Calculation on Lithium Hydride for Undergraduates†

Frank Rioux
Department of Chemistry, St. John's University, Collegeville, Minnesota 56321

Donald K. Harriss
Department of Chemistry, University of Minnesota—Duluth, Duluth, Minnesota 55812
(Received 2 July 1979; accepted 30 November 1979)

Students generally acquire an understanding of theoretical concepts only after they have attempted actual calculations based on those concepts. Consequently there is a need for realistic calculations in quantum theory which can be used with undergraduates to illustrate its important concepts. In this paper we describe a self-consistent-field–linear combination of atomic ortibals–molecular orbital calculation on the valence electrons of LiH using the method of Roothaan. The calculation is modest in scope, mathematically simple, and easy to program. These factors enhance its usefullness with undergraduates.

The recent literature in physics and chemistry contains several interesting examples of quantum-mechanical calculations for undergraduates.[1-5] These calculations are extremely useful in augmenting classroom work on quantum theory with "laboratory-like" experiences. They provide students with an opportunity to "do quantum mechanics," which increases the likelihood that they will master the associated formalism and conceptual framework. In addition, the exercises bring the students to the computer confronting them with many important numerical methods.

The student exercise presented here is a self-consistent-field–linear combination of atomic orbitals–molecular orbital (SCF-LCAO-MO) calculation on the valence electrons of LiH using the method of Roothaan,[6,7] an iterative procedure which yields a self-consistent set of wave functions and energies. The calculation is based on the work of Karo and Olsen[8,9] and was used by Murrell, Kettle, and Tedder[10] to illustrate SCF calculations. Given the importance of Roothaan's method in contemporary theoretical work, we thought that a more detailed discussion of the LiH calculation was justified.

SELF-CONSISTENT-FIELD THEORY

Using the fixed nuclei approximation and assuming that the $1s$ electrons of Li are not involved in bonding, reduces the LiH calculation to a two-electron (the classical valence electrons) problem. In the SCF treatment of the valence electrons, the two-electron Schrödinger equation is approximated by two one-electron equations of the form

$$(H_i^{\text{eff}} - \epsilon_i)\psi_i = 0, \quad i = 1,2, \tag{1}$$

where H_i^{eff}, the one-electron effective Hamiltonian is, in atomic units,

$$H_i^{\text{eff}} = -\tfrac{1}{2}\nabla_i^2 - \frac{1}{r_{\text{H}}} - \frac{3}{r_{\text{Li}}} + 2\int \frac{1}{r_{i,1s}}\phi_{s'}^2 d\tau_{s'} + \int \frac{1}{r_{ij}}\psi_j^2 d\tau_j. \tag{2}$$

The first term on the right-hand side is the kinetic energy operator for the ith electron, the second term represents its interaction with the hydrogen nucleus, and the third term is its interaction with the lithium nucleus. The fourth term is the average interaction of the ith valence electron with the two $1s$ electrons of lithium and the last term represents its average interaction with the other (jth) valence electron.

LINEAR COMBINATION OF ATOMIC ORBITALS–MOLECULAR ORBITAL

In an SCF calculation, the effective Hamiltonian cannot be determined until a trial wave function is chosen. In the most commonly used version of molecular orbital theory, the wave function is formed as a linear combination of atomic orbitals. In this case the atomic orbitals chosen are the valence orbitals, namely, the $1s$ of hydrogen (ϕ_h) and the $2s$ (ϕ_s) and $2p\sigma(\phi_p)$ of lithium:

$$\psi = c_h\phi_h + c_s\phi_s + c_p\phi_p. \tag{3}$$

The effective Hamiltonian can now be written as

$$H_i^{\text{eff}} = -\tfrac{1}{2}\nabla_i^2 - \frac{1}{r_{\text{H}}} - \frac{3}{r_{\text{Li}}} + 2(s'|s') + c_h^2(h|h) + c_s^2(s|s) + c_p^2(p|p) + 2c_hc_s(h|s) + 2c_hc_p(h|p) + 2c_sc_p(s|p), \tag{4}$$

where

$$(s'|s') = (\phi_{s'}|(1/r_{i,1s})|\phi_{s'})$$

and, for instance,

$$(h|p) = (\phi_h|(1/r_{ij})|\phi_p)$$

Note that s′ represents the lithium $1s$ orbital, while s represents the lithium $2s$ orbital.

Equations (3) and (4) are substituted into (1). Multiplication on the left by each member of the basis set (3) in turn and integration over all space yields three simultaneous, linear, and homogeneous equations in the coefficients c_h, c_s, and c_p:

†Reprinted from *Am. J. Phys.* 1980 **48** 439–41.

$$c_h(H_{hh} - \epsilon S_{hh}) + c_s(H_{hs} - \epsilon S_{hs}) + c_p(H_{hp} - \epsilon S_{hp}) = 0$$
$$c_h(H_{hs} - \epsilon S_{hs}) + c_s(H_{ss} - \epsilon S_{ss}) + c_p(H_{sp} - \epsilon S_{sp}) = 0$$
$$c_h(H_{hp} - \epsilon S_{hp}) + c_s(H_{sp} - \epsilon S_{sp}) + c_p(H_{pp} - \epsilon S_{pp}) = 0. \quad (5)$$

The notation used in Eqs. (5) is,

$$H_{hh} = (\phi_h | H_i^{\text{eff}} | \phi_h). \quad (6)$$

$$S_{hs} = (\phi_h | \phi_s). \quad (7)$$

Substitution of H_i^{eff} into Eq. (6) and using $r' = r_{\text{H}}$ and $r = r_{\text{Li}}$ yields,

$$H_{hh} = (h|-\tfrac{1}{2}\nabla_i^2|h) - (h|(1/r')|h) - 3(h|(1/r)|h) + 2(hh|s's') + c_h^2(hh|hh) + c_s^2(hh|ss) + c_p^2(hh|pp) + 2c_hc_s(hh|hs) + 2c_hc_p(hh|hp) + 2c_sc_p(hh|sp). \quad (8)$$

For the two electron integrals we have, for example,

$$(hh|hp) = \int \int \phi_h(1)\phi_h(2)|(1/r_{12})| \times \phi_p(2)\phi_h(1)\, d\tau_1 d\tau_2. \quad (9)$$

The orbitals of electron 1 are written to the left of the vertical bar and the orbitals of electron 2 to the right. The remaining elements of the determinant are given in the Appendix.

SELF-CONSISTENT-FIELD METHOD

Before the SCF calculation can begin, the integrals in Eq. (8) and those in the Appendix must be known. This is the most difficult and time consuming aspect of a molecular SCF calculation. Fortunately, the paper by Karo and Olsen[8] contains values for these integrals for LiH at nine internuclear distances. The calculation described in this paper is for the equilibrium internuclear distance of 3.0 bohr.

With the preliminary discussion of SCF theory completed, the students are given the following outline of the SCF method;

(i) Obtain the values of the necessary integrals from Ref. 8.

(ii) Choose initial values for the coefficients c_h, c_s, and c_p.

(iii) Calculate H_{ij} and S_{ij}. (Actually the overlap integrals do not vary from one iteration to the next.)

(iv) Expand the determinant of Eqs. (5) and solve the resulting polynomial for the lowest energy root.

(v) Use this value of ϵ in Eqs. (5) to obtain new and improved values for the coefficients.

(vi) Normalize the coefficients and return to step iii. The coefficients are normalized by multiplication by the normalization constant which is

$$N = 1/(c_h^2 + c_s^2 + c_p^2 + 2c_hc_sS_{hs} + 2c_hc_pS_{hp} + 2c_sc_pS_{sp})^{1/2}.$$

This iterative technique is continued until the calculation is self-consistent. That is, until the energy and the coefficients remain invariant from one iteration to the next. The calculation described above takes approximately two four-hour lab periods. One period to write the program and a second period to proofread the program, debug it, and run it.

RESULTS AND DISCUSSION

Table I contains the results of the calculation at 3.0 bohr as described above. Eighteen iterations were required for the calculation to achieve self-consistency to three significant figures. The last line of the table contains the results of Karo's calculation.[9] It can be seen that this calculation is in good agreement with the more accurate work of Karo.

According to Koopmans' theorem[11] the negatives of the one-electron energies approximate the ionization energies of the molecule. This calculation yields a first ionization energy of 8.3 eV for LiH, which compares favorably with the experimental value of 8.0 eV.[12]

Including the core electrons in the $1s$ orbital on Li, the electron density in the LiH molecule is given by

$$p = 2(1s_{\text{Li}})^2 + 2(c_h^2\phi_h^2 + c_s^2\phi_s^2 + c_p^2\phi_p^2 + 2c_hc_s\phi_h\phi_s + 2c_hc_p\phi_h\phi_p + 2c_sc_p\phi_s\phi_p). \quad (10)$$

An estimate of the relative distribution of charge between Li and H can be obtained by dividing the overlap density ($\phi_h\phi_s$, $\phi_h\phi_p$, and $\phi_s\phi_p$) equally between the two orbitals.[13] For instance,

$$\phi_h\phi_s = \tfrac{1}{2}S_{hs}(\phi_h^2 + \phi_s^2). \quad (11)$$

This approximation yields the following ionic character for the LiH molecule $-\text{Li}^{+0.44}\text{H}^{-0.44}$. This result is consistent with the fact that hydrogen is more electronegative than lithium.

CONCLUSION

Theoretical concepts generally acquire meaning for students only after they have attempted actual calculations. The calculation outlined above is a suitable exercise for undergraduates and, therefore, should aid the student in gaining insight into Roothaan's SCF method. In addition it should serve as a useful introduction to the literature in this area.

Table I. Summary of SCF orbital energy calculation for valence electrons of LiH.

Iteration	c_h	c_s	c_p	$-\epsilon$ (Hartrees)
1	0.707	0.500	0.500	0.410
2	0.867	−0.144	0.112	0.275
3	0.662	−0.390	0.193	0.330
4	0.799	−0.232	0.141	0.292
5	0.711	−0.333	0.179	0.315
6	0.769	−0.266	0.155	0.299
7	0.731	−0.310	0.171	0.310
8	0.756	−0.281	0.161	0.303
9	0.740	−0.300	0.168	0.307
10	0.751	−0.288	0.164	0.304
11	0.744	−0.296	0.166	0.306
12	0.748	−0.291	0.164	0.305
13	0.745	−0.294	0.166	0.306
14	0.747	−0.292	0.165	0.305
15	0.746	−0.293	0.166	0.306
16	0.747	−0.292	0.165	0.305
17	0.746	−0.293	0.165	0.305
18	0.746	−0.293	0.165	0.305
⋮	⋮	⋮	⋮	⋮
30	0.746	−0.293	0.165	0.305342
Ref. 9	(0.702)	(−0.329)	(0.205)	(0.305452)

APPENDIX

The remaining elements, H_{ij}, of Eqs. (5) are:

$$H_{ss} = (s| -\tfrac{1}{2}\nabla^2|s) - (s|(1/r')|s) - 3(s|(1/r)|s) + 2(ss|s's') + c_h^2(ss|hh) + c_s^2(ss|ss) + c_p^2(ss|pp) + 2c_hc_s(ss|hs) + 2c_hc_p(ss|hp) + 2c_2c_p(ss|sp);$$

$$H_{pp} = (p| -\tfrac{1}{2}\nabla^2|p) - (p|(1/r')|p) - 3(p|(1/r)|p) + 2(pp|s's') + c_h^2(pp|hh) + c_s^2(pp|ss) + c_p^2(pp|pp) + 2c_hc_s(pp|hs) + 2c_hc_p(pp|hp) + 2c_sc_p(pp|sp);$$

$$H_{hs} + (h| -\tfrac{1}{2}\nabla^2|s) - (h|(1/r')|s) - 3(h|(1/r)|s) + 2(hs|s's') + c_h^2(hs|hh) + c_s^2(hs|ss) + c_p^2(hs|pp) + 2c_hc_s(hs|hs) + 2c_hc_p(hs|hp) + 2c_sc_p(hs|sp);$$

$$H_{hp} = (h| -\tfrac{1}{2}\nabla^2|p) - (h|(1/r')|p) - 3(h|(1/r)|p) + 2(hp|s's') + c_h^2(hp|hh) + c_s^2(hp|ss) + c_p^2(hp|pp) + 2c_hc_s(hp|hs) + 2c_hc_p(hp|hp) + 2c_sc_p(hp|sp);$$

$$H_{sp} = (s| -\tfrac{1}{2}\nabla^2|p) - (s|(1/r')|p) - 3(s|(1/r)|p) + 2(sp|s's') + c_h^2(sp|hh) + c_s^2(sp|ss) + c_p^2(sp|pp) + 2c_hc_s(sp|hs) + 2c_hc_p(sp|hp) + 2c_sc_p(sp|sp).$$

[1] J. S. Bolemon, Am. J. Phys. **40**, 1511 (1972).
[2] J. S. Bolemon and D. J. Etzold, Jr., Am. J. Phys. **42**, 33 (1974).
[3] R. L. Snows and J. L. Bills, J. Chem. Educ. **52**, 506 (1975).
[4] R. T. Robiscoe, Am. J. Phys. **43**, 538 (1975).
[5] H. E. Montgomery, Jr., J. Chem. Educ. **54**, 742 (1977).
[6] C. C. J. Roothaan, Revs. Mod. Phys. **23**, 69 (1951).
[7] S. M. Blinder, Am. J. Phys. **33**, 431 (1965).
[8] A. M. Karo and A. R. Olsen, J. Chem. Phys. **30**, 1232 (1959).
[9] A. M. Karo, J. Chem. Phys. **30**, 1241 (1959).
[10] J. N. Murrell, S. F. A. Kettle, and J. M. Tedder, *Valence Theory* (Wiley, London, 1970), pp. 154–158.
[11] T. A. Koopmans, Physica **1**, 104 (1933).
[12] R. Velasco, Can. J. Phys. **35**, 1204 (1959).
[13] R. S. Mulliken, J. Chem. Phys. **23**, 1833, 1841 (1955).

Chapter 5

Energy Levels as Eigenvalues

5.1 Introduction

In the present chapter we shall be concerned with models in which the experimental data determine eigenvalues of an energy matrix. The models express the elements of the energy matrix linearly in terms of model parameters, the matrix coefficients of which describe the structure of the model. The usual aim is to determine parameters from experimental data and, in favourable situations where the parameters are over-determined, to test the model by the quality of fit. A second test of the model, which can be used even when the number of parameters and input data are equal, is to compare the parameter values obtained for similar systems.

We discuss two types of model in this chapter. The first is the *crystal-field model*, which is commonly used to describe the fine splittings in the optical spectra of paramagnetic ions in solids. This will be applied to lanthanide ions, for which the crystal-field split levels are characteristically sharp and provide excellent input data. The second is the '*harmonic*' *model* of molecular vibrations in which the restoring forces are proportional to atomic displacements (i.e. 'Hooke's law' forces). The vibrational frequencies required for input data are usually obtained by means of infrared spectroscopy. In both cases, the models are based on explicit hypotheses and provide good examples of 'falsifiable' phenomenological models.

5.2 Fitting Parameters to Eigenvalues

Spectroscopic measurements are commonly used to determine the energy differences between the stationary states of physical systems. In the absence of energy losses to coupled systems, sharply defined line spectra are obtained. For example, the early development of quantum mechanics was largely the result of the ease of obtaining good optical spectroscopic data for free atoms, especially simple atoms such as sodium in which the optical transitions can be explained in terms of the excitations of a single electron. We shall be interested in far more complicated systems which are not amenable to analysis from first principles but which may nevertheless be readily related to parametrised models.

For the purpose of this section we shall suppose that the raw spectroscopic data, which give energy differences between pairs of energy levels, have been analysed to obtain an empirical set of energy levels for the system under study. It is apparent from the nature of the data that the energy zero is undefined, so it is customary to set the lowest lying, or 'ground' state, at zero energy. Our aim is to relate the set of N measured energy levels to the eigenvalues of a model Hamiltonian, or force matrix, $\mathbf{H}$, which has fewer than N undetermined parameters. The quality of the fit obtained will provide one criterion for the acceptability of the model. Parameter values and model system eigenvectors may provide other criteria. There is little in the literature on fitting data to eigenvalues, but Cowan (1981, Chapter 16) provides a useful account.

In order to formulate this problem mathematically, we express the Hamiltonian or force matrix in the form

$$\mathbf{H} = \sum_k \mathbf{B}_k \theta_k \tag{5.1}$$

where θ_k are the undetermined parameters and the matrices $\mathbf{B}_k$ are fixed 'structure matrices' determined by the model. They are the analogues of the functions $f_\alpha(x)$ used in linear least squares fitting (see Chapter 1).

The computational problem is to simultaneously diagonalise the matrix $\mathbf{H}$ and determine the parameters θ_k which bring its eigenvalues into the closest coincidence with the diagonal matrix of empirical eigenvalues, $\mathbf{E}$. According to the analyses given in §§ (5.3) and (5.4), the diagonal elements of $\mathbf{E}$ may be squared vibration frequencies if $\mathbf{H}$ is a force matrix, or energy levels if $\mathbf{H}$ is the crystal-field Hamiltonian matrix. In algebraic terms, we have to minimise the error matrix

$$\boldsymbol{\varepsilon} = \mathbf{E} - \mathbf{U}^{\mathrm{T}}\mathbf{H}\mathbf{U} \tag{5.2}$$

where $\mathbf{U}$ is an orthogonal matrix ($\mathbf{U}^{\mathrm{T}}\mathbf{U} = \mathbf{I}$) chosen to diagonalise $\mathbf{H}$ *and* put its diagonal elements in an order corresponding to that of the elements of $\mathbf{E}$. In the absence of other information, we may order the diagonal elements in both matrices in decreasing order of magnitude (say). Other methods which may be used to deduce the correspondence between eigenvalues of $\mathbf{H}$ and the elements of $\mathbf{E}$ will be mentioned later in relation to specific examples.

As the matrix $\mathbf{U}$ will, in general, depend on the θ_k values, the minimisation of $\boldsymbol{\varepsilon}$ cannot be formulated as a linear least squares procedure and must therefore be carried out iteratively. A possible procedure is given below.

(i) Guess a starting set of parameter values $\theta_k (= \theta_k^{(1)})$ which then define a starting matrix $\mathbf{H}^{(1)}$. An informed guess is obviously preferable, such as might be based on a knowledge of the parameter values for similar systems.

(ii) Find the orthogonal matrix $\mathbf{U}^{(1)}$ which diagonalises $\mathbf{H}^{(1)}$. This is equivalent to the problem of finding all the eigenvectors of $\mathbf{H}^{(1)}$, as the columns of $\mathbf{U}^{(1)}$ are the (normalised) eigenvectors of $\mathbf{H}^{(1)}$.

(iii) Carry out a linear least squares fit of the parameters θ_k to the data matrix $\mathbf{E}$, while keeping the matrix $\mathbf{U}^{(1)}$ fixed. This problem is linear in the θ_k as we can write

$$\boldsymbol{\varepsilon} = \mathbf{E} - \sum_k \theta_k \mathbf{U}^{(1)\mathrm{T}} \mathbf{B}_k \mathbf{U}^{(1)} \tag{5.3}$$

where the fixed matrices $\mathbf{B}_k$ are those introduced in equation (5.1).

(iv) The set of parameters $\theta_k (= \theta_k^{(2)})$ determined in step (iii) are then used again in (i) and the cycle is repeated until the values $\theta_k^{(r)}$ converge.

In the above discussion we have been deliberately vague about the expression that should be minimised in the least squares fitting procedure. It is customary to minimise

$$D = \sum_i (\varepsilon_{ii})^2/\sigma_i^2 \tag{5.4}$$

where the σ_i are errors associated with the E_{ii}. The minimisation follows precisely the same pattern as the solution of equation (1.25) in Chapter 1, with $y_i = E_{ii}$ and the elements of the matrix $\mathbf{A}$ given by

$$(\mathbf{A})_{ik} = f_k(i) = (\mathbf{U}^{(r)\mathrm{T}} \mathbf{B}_k \mathbf{U}^{(r)})_{ii}$$

at the rth iteration.

The use of equation (5.4) to define the least squares procedure apparently makes good sense as, according to equation (5.2), the error matrix $\boldsymbol{\varepsilon}$ is diagonal by definition. Nevertheless, the $\mathbf{B}_k$ matrices in equation (5.3) are not, in general, separately diagonal, so the minimisation procedure in step (iii) may result in the creation of significant off-diagonal elements. It has been suggested (Newman 1981) that this problem can be avoided by minimising the alternative expression

$$D' = \mathrm{Tr}(\mathbf{W}^{1/2} \boldsymbol{\varepsilon} \mathbf{W}^{1/2} \boldsymbol{\varepsilon}) \tag{5.5}$$

where $\mathbf{W}$ is the weight matrix defined by equation (1.23) in Chapter 1.

An important property of equation (5.5) is that, by analogy with the use of orthogonal functions in linear least squares fitting, it is possible to define sets of mutually orthonormal matrices $\mathbf{B}_k$ which (in the case of real symmetric matrices) satisfy

$$\mathrm{Tr}(\mathbf{B}_k \mathbf{B}_m) = \delta_{km}. \tag{5.6}$$

In particular, we note that this relation is preserved under arbitrary orthogonal transformations of the matrices $\mathbf{B}_k$;

$$\mathbf{B}_k' = \mathbf{U}^{\mathrm{T}} \mathbf{B}_k \mathbf{U}$$

where $\mathbf{U}^{\mathrm{T}}\mathbf{U} = \mathbf{U}\mathbf{U}^{\mathrm{T}} = \mathbf{I}$. We then have

$$\mathrm{Tr}(\mathbf{U}^{\mathrm{T}} \mathbf{B}_k \mathbf{U} \mathbf{U}^{\mathrm{T}} \mathbf{B}_m \mathbf{U}) = \mathrm{Tr}(\mathbf{U}\mathbf{U}^{\mathrm{T}} \mathbf{B}_k \mathbf{U}\mathbf{U}^{\mathrm{T}} \mathbf{B}_m) = \mathrm{Tr}(\mathbf{B}_k \mathbf{B}_m) = \delta_{km}.$$

Applying the minimisation condition $\partial D'/\partial\theta_k = 0$ to equation (5.5) (with $\mathbf{W}$ equal to the unit matrix) and using equation (5.6), it can easily be shown that the fitting equations give the simple result

$$\theta_k = \mathrm{Tr}(\mathbf{U}^{(r-1)\mathrm{T}}\mathbf{B}_k\mathbf{U}^{(r-1)}\mathbf{E}). \tag{5.7}$$

As we shall find, in some physical problems the model determines a natural set of mutually orthonormal matrices $\mathbf{B}_k$.

Exercise

5.1 Before tackling the physically significant projects in parameter-fitting which follow, it is of interest to study the properties of the alternative fitting procedures corresponding to the minimisation of D and D'. This may be treated as a purely computational problem. An arbitrary diagonal data matrix may be selected, say

$$\mathbf{E} = \begin{bmatrix} 29 & & & & & \\ & 18 & & & 0 & \\ & & 7 & & & \\ & & & -3 & & \\ & 0 & & & -6 & \\ & & & & & -40 \end{bmatrix}$$

and the fit made to an arbitrarily transformed matrix of the form of (5.1), where the $\mathbf{B}_k$ values are mutually orthogonal and fewer in number than the dimensions of the data matrix. As an example, the reader may like to try fitting to the following set of four, mutually orthonormal, matrices:

$$\mathbf{B}_1 = \frac{1}{6}\begin{bmatrix} 1 & & & & & \\ & 1 & & & & \\ & & 1 & & & \\ & & & 1 & & \\ & & & & 1 & \\ & & & & & 1 \end{bmatrix} \qquad \mathbf{B}_3 = \frac{1}{\sqrt{10}}\begin{bmatrix} 0 & 1 & & & & \\ 1 & 0 & 1 & & & \\ & 1 & 0 & 1 & & \\ & & 1 & 0 & 1 & \\ & & & 1 & 0 & 1 \\ & & & & 1 & 0 \end{bmatrix}$$

$$\mathbf{B}_2 = \frac{1}{6}\begin{bmatrix} 1 & & & & & \\ & 1 & & & & \\ & & 1 & & & \\ & & & -1 & & \\ & & & & -1 & \\ & & & & & -1 \end{bmatrix} \qquad \mathbf{B}_4 = \frac{1}{\sqrt{8}}\begin{bmatrix} 0 & 0 & 1 & & & \\ 0 & 0 & 0 & 1 & & \\ 1 & 0 & 0 & 0 & 1 & \\ & 1 & 0 & 0 & 0 & 1 \\ & & 1 & 0 & 0 & 0 \\ & & & 1 & 0 & 0 \end{bmatrix}$$

Note that $\theta_3 = \theta_4 = 0$ should not be assumed in the starting set of parameters, as the iterative procedure given above can never change $\mathbf{U}$ from the value $\mathbf{U} = \mathbf{I}$ in this case. Several starting values for the parameters should be tried to see whether the fitting space has more than one minimum; one approach would be to take $\theta_k = \pm 5$ in various combinations. It is suggested that the alternative procedures corresponding to the minimisation of D and D' be compared in relation to their speed of convergence. The effect on θ_1, θ_2 and θ_3 of omitting $\mathbf{B}_4$ should also be tested.

5.3 The Crystal-field Parametrisation

In solids, paramagnetic ions often show a 'fine structure' in their optical spectrum due to the interaction of the electronic states of their partially filled shells with the crystalline environment. This interaction separates states which would otherwise have degenerate energy levels in the free ion into several distinct components, giving rise to the phenomenon known as 'crystal-field splitting' (see, for example, Hüfner (1978) for a more detailed discussion).

The superposition model of the crystal field has already been discussed in §2.4. In this section we describe the method of relating spectroscopic data, which determine the crystal-field split energy levels in paramagnetic ions, to phenomenological crystal-field parameters. As was mentioned in §2.4, it is not necessary to understand the origin of the crystal field in order to set up a phenomenological parametrisation. The crystal-field model is based on the assumption (not always explicit), that the crystal field acts independently on each electron in the open shell of the paramagnetic ion, i.e. that it is a one-electron field. Hence, the main theoretical problem consists of relating the many-electron energy matrix elements to the one-electron matrix elements of the crystal field. This involves a fairly complicated application of angular momentum theory which we shall by-pass by using results quoted in the literature and so move directly to the computational and physical aspects of the problem.

The free-ion states, which will be used as a basis set in our calculations, are characterised by a total angular momentum label L, a total spin label S and, if the spin–orbit coupling is stronger than the crystal field, by a total angular momentum label J as well. It is conventional to use spectroscopic notation to describe these states. The following letter codes substitute for the numerical values of L:

$L =$ 0	1	2	3	4	5	6	7	8	etc
$[L] =$ S	P	D	F	G	H	I	K	L	etc.

The three angular momenta are then grouped in the expression ${}^{(2S+1)}[L]_J$ so that, for example, we write ${}^5\mathrm{I}_8$ for a state with $S = 2$, $L = 6$ and $J = 8$. In Project 5A we shall only be considering the lowest lying, or *ground*, multiplets with a given value of J. These invariably satisfy the relation $J = L \pm S$, in conformity with Hund's rules. The ground multiplets of several lanthanide ions are studied in order to exemplify the methods of fitting parameters to eigenvalue data.

More generally, it is common to use optical spectroscopic values of the energy levels in many multiplets in such fittings, producing a considerable overdetermination of the crystal-field parameters, and hence providing a stringent test of the one-electron crystal-field model. In such fits, mean square deviations are frequently obtained which are only 2 to 3% of the mean splitting, showing that the major contributions to the observed crystal-field splitting can be attributed to the independent electron interactions with the environment which are included in the one-electron crystal-field model. When a crystal-field fit is carried out to many *LSJ* multiplets, it is now common practice to simultaneously fit 'free-ion' parameters, which describe isotropic effects, such as the Coulomb interaction between electrons in the open-shell and spin–orbit coupling. This requires the construction of very large energy matrices and it is beyond the scope of this work to describe the intricate mathematical details. Those readers who wish to pursue this topic further are referred to Abragam and Bleaney (1970) or Griffith (1961) for $3d^n$ ions, Hüfner (1978) for $4f^n$ ions and Cowan (1981) for a detailed discussion of fitting the isotropic (or free-ion) interactions.

The crystal-field Hamiltonian is usually written in operator form

$$H = \sum_{nm} B_m^n O_m^{(n)} \tag{5.8}$$

where the so-called 'tensor' operators, $O_m^{(n)}$, act on the single-electron states. The Wigner–Eckart theorem (Elliott and Dawber 1979, §4.20) allows us to write the $O_m^{(n)}$ as proportional to operators which act directly on the total angular momentum states, and hence can be expressed as functions of the total angular momentum operator J. Therefore, we write

$$O_m^{(n)} = \gamma_n O_m^{(n)}(J) \tag{5.9}$$

where the factors γ_n (which according to the Wigner–Eckart theorem are independent of m) depend on the free-ion states and have been tabulated for many systems of interest. This method of replacing $O_m^{(n)}$ by $O_m^{(n)}(J)$ is referred to as 'the method of operator equivalents' in the literature and was formulated by Stevens (1952). It is conceptually simpler than the explicitly group-theoretical methods, although it must be recognised that the underlying Wigner–Eckart theorem is itself a group-theoretical result.

The first step of the computational problem is thus resolved into constructing the energy matrix by evaluating H for the $2J + 1$ states $|LSJM_J\rangle$ of the *LSJ* ground multiplet labelled by M_J. As the J multiplets are degenerate until the crystal-field perturbation is 'switched on', the eigenvalues of this energy matrix should correspond to the observed energy levels. Hence, the second step is to carry out a least squares fitting procedure to determine the best values of the parameters B_m^n to fit the data. Further details are discussed in the description of Project 5A, which follows.

Exercise

5.2 Show that the four matrices $O_m^{(n)}(J)$ for $J = 9/2$ and $J = 6$ (table 5.4) are mutually orthogonal in the sense that $\mathrm{Tr}(O_m^{(n)}(J)O_{m'}^{(n')}(J)) = 0$ if $n \neq n'$ or $m \neq m'$. In the case where $J = 9/2$, use this result to establish a direct relationship between the energy levels and crystal-field parameters.

PROJECT 5A: FITTING CRYSTAL-FIELD PARAMETERS TO ENERGY LEVELS

The aim of the project is to apply the fitting procedures described in §5.2 to determine the 'best fit' parameters for the experimental energy levels given in table 5.1. It will be found that these parameters are not, in general, exactly the same as the published parameter sets quoted in table 5.2. There are two reasons for this. It is possible that the values of the factors γ_n (see equation (5.9)) used in deriving the published parameter values may be different to those we give in table 5.3. Also, perhaps more significantly, the published parameters have been derived as the result of fitting to many more energy levels than those given in table 5.2. The deviations between parameters fitted (in this project) to just the ground multiplet levels and the published parameters may thus be taken to provide an indication of the imperfections of the one-electron crystal field model.

It is best to approach the computational aspect of this project in two stages, as indicated at the end of §5.3. The first stage requires only a program to diagonalise the energy matrices, while the second stage will also require the use of the least squares fitting procedures described in §5.2. Energy matrices for the one-electron operators $O_m^{(n)}$ may be constructed using equations (5.8) and (5.9), together with the values tabulated in tables 5.3 and 5.4.

In our examples we shall only require matrix elements of $O_m^{(n)}(J)$ for (n, m) equal to (2, 0), (4, 0), (6, 0) and (6, 6). These are given in table 5.4. More complete tables of matrix elements of these operators can be found in the book by Abragam and Bleaney (1970) and in the review article by Hutchings (1964).

The first stage of this project consists of constructing the energy matrix and solving the eigenvalue problem for some specific matrices of the crystal-field Hamiltonian H in equation (5.8). Library programs, or a program developed by the reader on the lines suggested in Chapter 4, can be used for this purpose. Systems with crystal structure $R(C_2H_5SO_4) \cdot 9H_2O$ and RCl_3 (where R denotes a lanthanide ion) have been chosen as examples because the high symmetry of the lanthanide site is reflected in the need to employ only four crystal-field parameters—B_0^2, B_0^4, B_0^6, B_6^6. Group-theoretical methods (Elliott and Dawber 1979, Leech and Newman 1969, Cotton 1963) can then be used to block-diagonalise the matrix of H, which simplifies the computational problem considerably. We do not suggest the employment of

Table 5.1 Experimental energy levels (cm^{-1}). The structure of the eigenstates in terms of M_J level admixtures is indicated by the bracketed numbers. These provide a means of identifying corresponding fitted energy levels.

$^4I_{9/2}$, $Nd(C_2H_5SO_4)_3 9H_2O$†		$^5I_8$, Ho:$LaCl_3$‡		$^4I_{9/2}$, Nd:$LaCl_3$§		$^6H_{15/2}$, Dy:$LaCl_3$‖		$^7F_6$, $Tb(OH)_3$¶	
0	($\pm 7/2$, $\mp 5/2$)	0	(-5, 1, 7)	0	($\pm 7/2$, $\mp 5/2$)	0	($\pm 3/2$, $\mp 9/2$, $\pm 15/2$)	0	(0, 6)
149.4	($\pm 1/2$)	12.51	(6, 0, -6)	115.39	($\pm 1/2$)	9.82	($\pm 5/2$, $\mp 7/2$)	118.2	(1, -5)
154	($\pm 9/2$, $\mp 3/2$)	43.80	(6, 0, -6)	123.21	($\pm 9/2$, $\mp 3/2$)	9.97	($\pm 3/2$, $\mp 9/2$, $\pm 15/2$)	206.0	(2, -4)
279	($\pm 7/2$, $\mp 5/2$)	66.42	(-5, 1, 7)	244.40	($\pm 7/2$, $\mp 5/2$)	15.65	($\pm 1/2$, $\mp 11/2$, $\pm 13/2$)	224.4	(0, 6)
311	($\pm 9/2$, $\mp 3/2$)	89.92	(-4, 2, 8)	249.35	($\pm 3/2$, $\mp 9/2$)	40.75	($\pm 1/2$, $\mp 11/2$, $\pm 13/2$)	233.7	(1, -5)
		104.12	(3)			80.48	($\pm 5/2$, $\mp 7/2$)	?	(3)
		118.40	(6, 0, -6)			121.65	($\pm 3/2$, $\mp 9/2$, $\pm 15/2$)	252.3	(2, -4)
		154.21	(-4, 2, 8)			140.51	($\pm 1/2$, $\mp 11/2$, $\pm 13/2$)	?	(3)
		155.41	(-5, 1, 7)						
		203.69	(-4, 2, 8)						
		212.78	(3)						

†Gruber and Satten (1963).
‡Dieke and Pandey (1964).
§Carlson and Dieke (1961).
‖Crosswhite and Dieke (1961).
¶Scott *et al* (1969).

Table 5.2 Crystal-field parameters for some trivalent lanthanide ions in the anhydrous chlorides† (all in cm^{-1}).

	B_0^2	B_0^4	B_0^6	B_6^6	Reference
Nd^{3+}:$LaCl_3$	82	−42	−45	438	Carlson and Dieke (1961)
Ho^{3+}:$LaCl_3$	114	−34	−28	277	Dieke and Pandey (1964)
Dy^{3+}:$LaCl_3$	97	−41	−29	272	Crosswhite and Dieke (1961)

†A more complete listing will be found in Hüfner (1978).

such techniques in the present case however, because the matrices are quite small and can, in any case, be partially block-diagonalised by inspection. This can be seen from the sets of M_J values given in table 5.1, which show that only a few elements of the M_J basis are involved in a given eigenvector.

In making comparisons between your calculated eigenvalues and the energy levels given in table 5.1 it is necessary to keep the following points in mind.

(i) The energy zero of the tabulated levels is arbitrary. You will find that the *mean* energy of levels determined using H is zero, rather than this being the energy of the lowest-lying state as is conventionally assumed in table 5.1. This can be dealt with either by shifting the input energies or by introducing an additional parameter corresponding to the mean energy (B_0^0).

(ii) The matrix of H only mixes together a small number of different M_J values. Each diagonal block of H, produced by permuting rows and columns, will correspond to a well defined set of M_J values, facilitating comparison with the empirical results in table 5.1, where M_J admixtures are identified.

(iii) Degenerate energy levels appear more than once as a result of diagonalising the matrix of H. It should be possible to eliminate repeated (block diagonal) submatrices from the calculation by inspection, and this should be incorporated into the data input process. In particular, it should be noted that, when the J values are half-integral, all energy levels are at least doubly degenerate. In the case of Nd^{3+}, for example, the ground state has $J = 9/2$ and multiplicity $(2J + 1) = 10$. Only five energies are listed, however, because each is doubly degenerate.

In the second stage of the project it will be necessary to incorporate the least squares procedure described in §5.2. We can then determine the best fit crystal-field parameters corresponding to the energy levels given in table 5.1. In particular, it should be possible to estimate the $Tb(OH)_3$ energy levels which were not determined experimentally (marked by '?' in table 5.1).

Table 5.3 Values of the coefficients γ_n†.

	Nd^{3+}, $^4I_{9/2}$	Ho^{3+}, $^5I_8$	Dy^{3+}, $^6H_{15/2}$	Tb^{3+}, $^7F_6$
γ_2	$-7/(3^2 \times 11^2)$	$-1/(2 \times 3^2 \times 5^2)$	$-2/(3^2 \times 5 \times 7)$	$-1/(3^2 \times 11)$
γ_4	$-2^3 \times 17/(3^3 \times 11^3 \times 13)$	$-1/(2 \times 3 \times 5 \times 7 \times 11 \times 13)$	$-2^3/(3^2 \times 5 \times 7 \times 11 \times 13)$	$2/(3^3 \times 5 \times 11^2)$
γ_6	$-5 \times 17 \times 19/(3^3 \times 7 \times 11^3 \times 13^2)$	$-5/(3^3 \times 7 \times 11^2 \times 13^2)$	$2^2/(3^3 \times 7 \times 11^2 \times 13^2)$	$-1/(3^4 \times 7 \times 11^2 \times 13)$

†Also, see Hüfner (1978), table 8.

Table 5.4 Matrix elements of the operators $O_m^{(n)}(J)$. The F numbers in the second column are multiplying factors common to all elements in a row. Expressions for $O_0^{(2)}$ etc are given at the end of the table.

Operator	J	F	$M_J = \pm 1/2$	$\pm 3/2$	$\pm 5/2$	$\pm 7/2$	$\pm 9/2$	$\pm 11/2$	$\pm 13/2$	$\pm 15/2$
$O_0^{(2)}$	9/2	6	− 4	− 3	− 1	2	6			
	15/2	3	− 21	− 19	− 15	− 9	− 1	9	21	35
$O_0^{(4)}$	9/2	84	18	3	− 17	− 22	18			
	15/2	60	189	129	23	− 101	− 201	− 221	− 91	273
$O_0^{(6)}$	9/2	5 040	− 8	6	10	− 11	3			
	15/2	13 860	− 75	− 25	45	87	59	− 39	− 117	65

Operator	J	F	$M_J = (\pm 7/2 \| \mp 5/2)$	$(\pm 9/2 \| \mp 1/2)$	$(\pm 11/2 \| \mp 1/2)$	$(\pm 13/2 \| \pm 1/2)$	$(\pm 15/2 \| \pm 3/2)$
$O_6^{(6)}$	9/2	360	14	$2\sqrt{21}$			
	15/2	$360\sqrt{11}$	$42\sqrt{5}$	84	$4\sqrt{273}$	$7\sqrt{39}$	$\sqrt{455}$

Operator	J	F	$M_J = 0$	± 1	± 2	± 3	± 4	± 5	± 6	± 7	± 8
$O_0^{(2)}$	6	3	− 14	− 13	− 10	− 5	2	11	22		
	8	3	− 24	− 23	− 20	− 15	− 8	1	12	25	40
$O_0^{(4)}$	6	60	84	64	11	− 54	− 96	− 66	99		
	8	420	36	31	17	3	− 24	− 39	− 39	− 13	52
$O_0^{(6)}$	6	7 560	− 40	− 20	22	43	8	− 55	22		
	8	13 860	− 120	− 85	2	93	128	65	− 78	− 169	104

Operator	J	F	$M_J = (3 \| -3)$	$(\pm 4 \| \mp 2)$	$(\pm 5 \| \mp 1)$	$(\pm 6 \| 0)$	$(\pm 7 \| \pm 1)$	$(\pm 8 \| \pm 2)$
$O_6^{(6)}$	6	360	84	$14\sqrt{30}$	$7\sqrt{66}$	$2\sqrt{231}$		
	8	360	462	$42\sqrt{110}$	$12\sqrt{1001}$	$14\sqrt{429}$	$7\sqrt{715}$	$2\sqrt{2002}$

$O_0^{(2)} = 3J_z^2 - J(J+1).$

$O_0^{(4)} = 35J_z^4 - 30J(J+1)J_z^2 + 25J_z^2 - 6J(J+1) + 3J^2(J+1)^2.$

$O_0^{(6)} = 231J_z^6 - 315J(J+1)J^4 + 735J_z^4 + 105J^2(J+1)^2J_z^2 - 525(J+1)J_z^2 + 294J_z^2 - 5J^3(J+1)^3 + 40J^2(J+1)^2 - 60J(J+1).$

$O_6^{(6)} = \frac{1}{2}(J_+^6 + J_-^6).$

5.4 Molecular Vibrations

A wide range of systems exist for which the resonant frequencies and normal modes of vibration can be obtained by diagonalising a finite matrix, the $n \times n$ elements of which give the coupling between the degrees of freedom of the system. Consider, for example, N point masses coupled so that the displacement of any one of them from its equilibrium position produces a restoring force due to its interactions with its neighbours. This is the situation in free molecules, where any displacement of the atoms from their equilibrium positions sets up a network of interactions tending to restore the atoms to their original relative positions. The simplest mathematical model for such systems is the harmonic approximation. It can be expressed by writing the restoring force $\boldsymbol{f}_\mathrm{k}$ on atom k as proportional to the displacements of all N atoms, namely

$$\boldsymbol{f}_\mathrm{k} = -\sum_{j=1}^{N} \mathbf{F}_\mathrm{kj} \cdot \boldsymbol{r}_\mathrm{j}. \tag{5.10}$$

Here, the sum is taken over all the atom displacement vectors $\boldsymbol{r}_\mathrm{j}$ and the $\mathbf{F}_\mathrm{kj}$ are 3×3 force constant matrices giving the coupling between atoms k and j. The minus sign is introduced for convenience, allowing the restoring forces $\mathbf{F}_\mathrm{kk}$ to be written as positive quantities. The dot product implies a summation over the three coordinate directions associated with the atom j.

Using Newton's second law, equation (5.10) leads at once to the dynamical equations

$$m_\mathrm{k}\ddot{\boldsymbol{r}}_\mathrm{k} = -\sum_{j=1}^{N} \mathbf{F}_\mathrm{kj} \cdot \boldsymbol{r}_\mathrm{j}. \tag{5.11}$$

These equations may also be written more simply in terms of a single $3N \times 3N$ matrix if we define a vector $\boldsymbol{R}$ with three-dimensional subvectors $\boldsymbol{R}_\mathrm{k}$ and a matrix $\mathbf{H}$ with 3×3 submatrices $\mathbf{H}_\mathrm{kj}$ such that

$$\boldsymbol{R}_\mathrm{k} = \sqrt{m_\mathrm{k}}\boldsymbol{r}_\mathrm{k} \qquad \mathbf{H}_\mathrm{kj} = \boldsymbol{F}_\mathrm{kj}/\sqrt{m_\mathrm{k} m_\mathrm{j}}.$$

The equations (5.11) are collected in a single matrix equation of the form

$$\ddot{\mathbf{R}} = -\mathbf{HR}. \tag{5.12}$$

Normal modes of vibration correspond to a motion in which all atoms oscillate coherently about their equilibrium positions with the same frequency, but not necessarily with similar amplitudes or phases. This corresponds to a solution in the form $\mathbf{R} = \mathbf{R}_0 \cos(\omega t + \delta)$, where $\mathbf{R}_0$ is time independent, and reduces equation (5.12) to the eigenvalue equation

$$(\mathbf{H} - \omega^2)\mathbf{R}_0 = 0. \tag{5.13}$$

The matrix $\mathbf{H}$ can easily be shown to be symmetric (as a consequence of action being equal to reaction) and to have positive, or zero, eigenvalues because of the requirement that the undistorted system is in equilibrium.

Molecular vibration frequencies are usually obtained by spectroscopic absorption or scattering measurements (see, for example, Banwell 1966). The mathematical aspects of the relation between the force constants and vibration frequencies have been studied in depth by Wilson *et al* (1980). Many books on the applications of group theory in physics and chemistry use molecular vibrations as an example (see Elliott and Dawber 1979, Chapter 6, Cotton 1963, Leech and Newman 1969 and Nussbaum 1971).

The construction of the force matrix $\mathbf{F} = [F_{ij}]$ is a necessary preliminary to finding the eigenfrequencies and normal modes of vibration of a molecule. Many different choices of coordinate system are possible, but physical chemists generally choose coordinates which are closely related to the internal constraints, such as changes in bond lengths and changes in the angles between bonds (for example, Wilson *et al* 1980 and Nussbaum 1971). These 'internal coordinates' automatically eliminate rigid molecule displacements but may involve linear dependence of the remaining coordinates. It is thus conceptually simpler to use cartesian coordinates, which is done in the following discussion of the linear molecule CO_2. In our discussion of CH_3Cl, however, we shall find it more convenient to use internal coordinates because this approach has so frequently been employed in the literature.

The problem is to express the matrices $\mathbf{F}_{ij}$ introduced in equation (5.10) as a linear expression in a set of force constants α_k

$$\mathbf{F}_{ij} = \sum_k \alpha_k \mathbf{F}_{ij}^{(k)} \tag{5.14}$$

where the $\mathbf{F}_{ij}^{(k)}$ are numerical coefficient matrices and the α_k are to be determined by least squares fitting as discussed in §5.2. This is trivial if internal coordinates are used but may be quite involved if cartesian coordinates are used for non-linear molecules because the use of rotation matrices is required to relate local coordinates aligned to the bonds with the common coordinates of the system. Having seen the details of the following example, readers may care to attempt to set up the cartesian coordinate force matrix for molecules with the CH_3Cl structure. They should be warned, however, that this problem is far from trivial.

Construction of the Force Matrix for CO_2

This construction is very simple because the molecule is linear, making it possible to express displacements directly in terms of cartesian coordinates. We shall employ the atom labels and coordinate directions shown in figure 5.1. In a three-atom system, two distinct types of restoring force can occur: that due to bond stretching, shown in figure 5.2, and that due to changes in the angle between neighbouring bonds, shown in figure 5.3. The complete

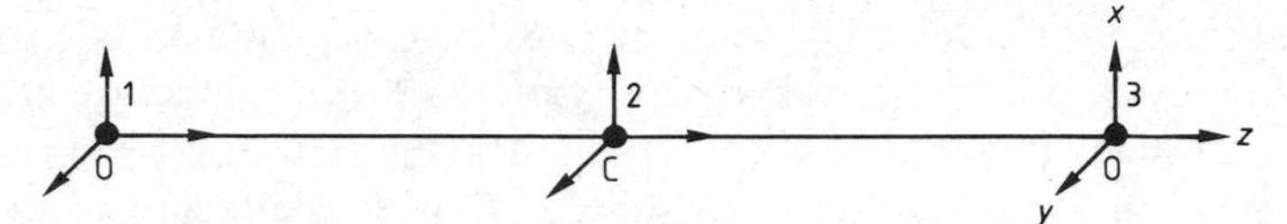

Figure 5.1 Alignment of CO_2 coordinate systems.

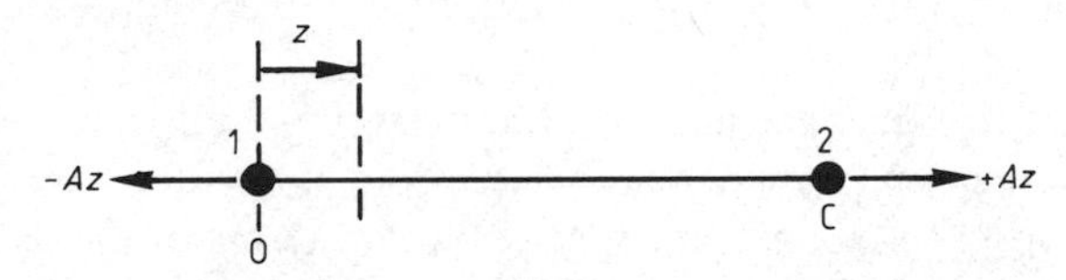

Figure 5.2 Forces due to a $+z_1$ displacement.

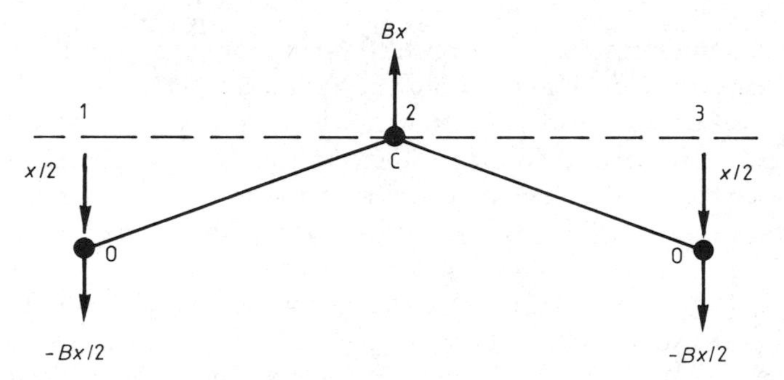

Figure 5.3 Bending mode forces due to $x_1/2$ and $x_3/2$ displacements.

force matrix, based on the simple procedure described below, is given by

$$
\mathbf{F} = \begin{array}{c} \\ 1 \\ \\ \\ 2 \\ \\ \\ 3 \\ \\ \end{array}
\left[\begin{array}{ccc|ccc|ccc}
x_1 & y_1 & z_1 & x_2 & y_2 & z_2 & x_3 & y_3 & z_3 \\
-\frac{B}{2} & 0 & 0 & B & 0 & 0 & -\frac{B}{2} & 0 & 0 \\
0 & -\frac{B}{2} & 0 & 0 & B & 0 & 0 & -\frac{B}{2} & 0 \\
0 & 0 & -A & 0 & 0 & A & 0 & 0 & 0 \\
\hline
B & 0 & 0 & -2B & 0 & 0 & B & 0 & 0 \\
0 & B & 0 & 0 & -2B & 0 & 0 & B & 0 \\
0 & 0 & A & 0 & 0 & -2A & 0 & 0 & A \\
\hline
-\frac{B}{2} & 0 & 0 & B & 0 & 0 & -\frac{B}{2} & 0 & 0 \\
0 & -\frac{B}{2} & 0 & 0 & B & 0 & 0 & -\frac{B}{2} & 0 \\
0 & 0 & 0 & 0 & 0 & A & 0 & 0 & -A
\end{array}\right]
$$

We now proceed to show how several columns of this matrix were derived.

Consider a displacement of atom 1 in the z direction, the other two atoms remaining fixed. The restoring force on atom 1 is in the z direction and equal to $-Az_1$. An equal and opposite force Az_1 acts on atom 2, but no force acts on atom 3. It follows that the z_1 column of **F** takes the form shown above. The z_3 column is obtained similarly. However, the z_2 column has a different form, for the motion of atom 2 in the z direction stretches both bonds and gives a restoring force of $-2Az_2$, balanced by forces Az_2 acting on both of the other atoms.

Now consider a displacement x_1. The resulting configuration of atoms is equivalent to the bond bending shown in figure 5.3, with $-x = x_1$. It follows that the restoring force on atom 1 is $-Bx_1/2$ in terms of the usual definition of the bond-bending constant B. The balancing forces on the other atoms are then determined in terms of B, establishing the form of the x_1 column of **F**.

A similar argument can be applied to obtain the y_1, x_3 and y_3 of columns of **F**. However, we obtain a different result for the x_2 column. Figure 5.3 shows that we must identify $x/2$, which equals x_2 in this case, and the restoring force is therefore $-2Bx_2$ on atom 2, counterbalanced by Bx_2 on atoms 1 and 3. The x_2 column of the force matrix thus takes the form shown above. The y_2 column is (owing to symmetry) completely analogous.

The final matrix can easily be checked for symmetry. It should also be noted that the sum of elements in any row or column vanishes, corresponding to the fact that there can be no net force on the whole molecule due to its internal motions.

Diagonalisation of this matrix is particularly easy as the x, y and z motions are not mutually coupled, so that the largest matrix to be treated is 3×3. The reader may care to check, either analytically or by computation, that the non-zero eigenvalues, ω, are given by $\omega^2 = (A/m_0)$, $A[(1/m_0) + (2/m_0)]$ and $B[(1/m_0) + (2/m_0)]$ (twice).

PROJECT 5B: FORCE CONSTANT MODEL OF THE VIBRATIONAL MODES OF CH_3Cl (METHYL CHLORIDE) AND CH_4 (METHANE)

Wilson *et al* (1980) and Colthup *et al* (1975) have discussed the vibrational modes of the molecule $CHCl_3$ (chloroform) using internal coordinates in some detail. This has the same structure as CH_3Cl but different bonding angles, so that it is trivial to adapt the algebraic results given in these works to the present problem. We shall not, however, follow the group-theoretical developments in those papers which allow the force and mass matrices to be block-diagonalised and hence significantly reduce the complexity of the computational problem.

The advantage of using the internal coordinates shown in figure 5.4 is that the coordinates corresponding to rigid translations and rotations are automatically eliminated and it is fairly easy to construct the force matrix. The disadvantages are that there are many force constants and relationships between them are not easy to establish, some of the internal coordinates may be superfluous and the kinetic energy expression takes a rather complicated form. In the present example, one of the ten internal coordinates is superfluous (as there are only nine internal degrees of freedom for five atoms). In practice, this is not a serious problem; it merely provides one zero eigenvalue of the dynamical equations. The algebraic form of the kinetic energy expression has been discussed, in general terms, by Wilson *et al* (1980, Appendix VI) and their results for a specific bond angle (109.47°) corresponding to tetrahedral coordination (i.e. CH_4 structure) are quoted in Nussbaum (1971) and Colthup *et al* (1975). We shall assume that the other molecules considered in this project are sufficiently close to tetrahedral for the same results to be used.

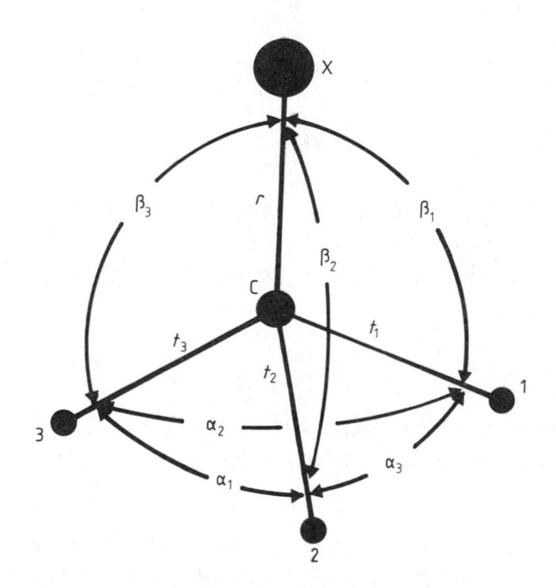

Figure 5.4 The CH_3X structure showing internal coordinates. The hydrogen positions are labelled 1, 2, 3. X is a halogen.

Adapting the expressions from Colthup *et al* (1975), the kinetic energy may be written as $T = \frac{1}{2}\dot{\boldsymbol{q}}^{\mathrm{T}}\mathbf{G}\dot{\boldsymbol{q}}$, where $\boldsymbol{q}$ is the vector of internal coordinates and $\mathbf{G}$, known as the kinetic energy matrix, may be expressed in terms of the internal coordinates shown in figure 5.4 ($i, j = 1, 2, 3$) as follows:

$$
\begin{aligned}
&G_{rr} = m_{\mathrm{C}} + m_{\mathrm{X}} \\
&G_{t_i, t_j} = G_{rt_i} = -\tfrac{1}{3}m_{\mathrm{C}} \qquad (i \neq j) \\
&G_{r\alpha_j} = \frac{2\sqrt{2}}{3}\frac{m_{\mathrm{C}}}{R_{\mathrm{H}}} \\
&G_{t_it_i} = m_{\mathrm{C}} + m_{\mathrm{H}} \\
&G_{t_i\alpha_i} = -G_{r\beta_i} = -G_{t_i\alpha_j} = \frac{2\sqrt{2}}{3}\frac{m_{\mathrm{C}}}{R_{\mathrm{H}}} \qquad (i \neq j) \\
&G_{t_i\beta_i} = -\frac{2\sqrt{2}}{3}\frac{m_{\mathrm{C}}}{R_{\mathrm{X}}} \qquad G_{t_i\beta_j} = \frac{\sqrt{2}}{3}\left(\frac{1}{R_{\mathrm{H}}} + \frac{1}{R_{\mathrm{X}}}\right)m_{\mathrm{C}} \qquad (i \neq j) \\
&G_{\alpha_i\alpha_i} = (\tfrac{8}{3}m_{\mathrm{C}} + 2m_{\mathrm{H}})/R_{\mathrm{H}}^2 \qquad G_{\alpha_i\alpha_j} = -m_{\mathrm{H}}/(2R_{\mathrm{H}}^2) \qquad (i \neq j) \\
&G_{\alpha_i\beta_i} = -\frac{4}{3}m_{\mathrm{C}}\left(\frac{1}{R_{\mathrm{H}}} + \frac{1}{R_{\mathrm{X}}}\right)\frac{1}{R_{\mathrm{H}}} \\
&G_{\alpha_i\beta_j} = -\frac{2m_{\mathrm{C}}}{3}\left(\frac{1}{R_{\mathrm{H}}} - \frac{1}{R_{\mathrm{X}}}\right)\frac{1}{R_{\mathrm{H}}} - \frac{m_{\mathrm{H}}}{2}\frac{1}{R_{\mathrm{H}}^2} \qquad (i \neq j) \\
&G_{\beta_i\beta_i} = \frac{m_{\mathrm{X}}}{R_{\mathrm{X}}^2} + \frac{m_{\mathrm{H}}}{R_{\mathrm{H}}^2} + m_{\mathrm{C}}\left(\frac{1}{R_{\mathrm{X}}^2} + \frac{1}{R_{\mathrm{H}}^2} + \frac{2}{3}\frac{1}{R_{\mathrm{H}}R_{\mathrm{X}}}\right) \\
&G_{\beta_i\beta_j} = -\frac{1}{2}\frac{m_{\mathrm{X}}}{R_{\mathrm{X}}^2} - \frac{m_{\mathrm{C}}}{6}\left(\frac{3}{R_{\mathrm{X}}^2} + \frac{2}{R_{\mathrm{H}}R_{\mathrm{X}}} - \frac{5}{R_{\mathrm{H}}^2}\right) \qquad (i \neq j).
\end{aligned}
\tag{5.15}
$$

In these equations R_{X} and R_{H} refer to the distances C–X and C–H respectively and $m_{\mathrm{H}}, m_{\mathrm{C}}, m_{\mathrm{X}}$ are the atomic masses.

The equations of motion (i.e. Newton's second law) may then be written in the form

$$
\frac{\mathrm{d}}{\mathrm{d}t}\left(\frac{\partial T}{\partial \dot{q}_\lambda}\right) + \frac{\partial V}{\partial q_\lambda} = 0 \tag{5.16}
$$

where the kinetic energy T is defined above, and the potential function V is related to the generalised force matrix $\mathbf{F}$ by

$$
V = \boldsymbol{q}^{\mathrm{T}}\mathbf{F}\boldsymbol{q}.
$$

In a completely general phenomenological model, there are a similar number of independent force matrix elements (16) to the number of elements shown above for the kinetic energy matrix. Not all of these can be regarded as being independent, however, if we wish to propose a model for the six observed vibrational frequencies. Given the aims of the present project, therefore, we shall need to use a much simpler force matrix. There are, of course, various ways of doing this, one obvious way being to retain only the four diagonal elements F_{rr}, $F_{t_it_i}$, $F_{\alpha_i\alpha_i}$, $F_{\beta_i\beta_i}$. The sets of force constants quoted in Nussbaum (1971, table 2.20) and Colthup *et al* (1975, table 14.11) may suggest alternative models. It should be noted that in the case of the CH_4

molecule $F_{rr} = F_{t_it_i}$ and $F_{\alpha_i\alpha_i} = F_{\beta_i\beta_i}$, so that there are only two diagonal force constants (and four observed frequencies).

The computational problems reduce to finding solutions of the eigenvalue equation which follows from equation (5.16)

$$(\mathbf{F} - \mathbf{G}\omega^2)\boldsymbol{\phi} = 0 \tag{5.17}$$

where $\mathbf{G}$ is the kinetic energy matrix given above and $\mathbf{F}$ is a (possibly diagonal) force matrix. The standard method of solving this equation, which exploits the fact that both $\mathbf{F}$ and $\mathbf{G}$ are symmetric, is to first solve the eigenvalue equation for $\mathbf{G}$

$$(\mathbf{G} - \lambda)\boldsymbol{u} = 0$$

and then use the orthogonal matrix $\mathbf{U} = \{\boldsymbol{u}_1, \boldsymbol{u}_2 \ldots\}$, constructed from the eigenvectors $\boldsymbol{u}_i$, to diagonalise $\mathbf{G}$

$$\boldsymbol{\Lambda} = \mathbf{U}^{\mathrm{T}}\mathbf{G}\mathbf{U}.$$

It is then possible to construct a square root of $\mathbf{G}$, $\mathbf{G}^{1/2} = \mathbf{U}\boldsymbol{\Lambda}^{1/2}\mathbf{U}^{\mathrm{T}}$, and its inverse $\mathbf{G}^{-1/2} = \mathbf{U}\boldsymbol{\Lambda}^{-1/2}\mathbf{U}^{\mathrm{T}}$, where $\boldsymbol{\Lambda}^{1/2}$ is taken to have diagonal elements which are positive square roots of the diagonal elements of $\boldsymbol{\Lambda}$. Then the original eigenvalue equation can be re-expressed in the form,

$$(\mathbf{G}^{-1/2}\mathbf{F}\mathbf{G}^{-1/2} - \omega^2)(\mathbf{G}^{1/2}\boldsymbol{\phi}) = 0. \tag{5.18}$$

In this case, therefore, it is only necessary to diagonalise the symmetric matrix $\mathbf{G}^{-1/2}\mathbf{F}\mathbf{G}^{1/2}$.

Data for the present project can be found in Chapter III of Herzberg (1945) and in tables 5.5 and 5.6. Starting values of the force constants for a non-linear least squares fit might be

$$F^0_{t_it_i} = F^0_{rr} = 4 \times 10^{-8}\ \mathrm{N\ nm}^{-1}$$
$$F^0_{\alpha\alpha} = F^0_{\beta\beta} = 1 \times 10^{-8}\ \mathrm{N\ nm}^{-1}.$$

In carrying out this project there are several criteria that can be used to check the effectiveness of the model. It should be possible to obtain an accurate fit to the data using fewer parameters than the given number of frequencies. A given parameter, say for H–C–H bending, should take very similar values for all the systems. Force constants for the systems involving H and D should be very similar. Your conclusions should be related to these criteria.

Table 5.5 Resonant frequencies of tetrahedral molecules (cm^{-1}) from Herzberg (1945).

	Degeneracy	CH_4	CD_4
ν_1	1	2914.2	2084.7
ν_2	2	1526 (approx.)	1054 (approx.)
ν_3	3	3020.3	2258.2
ν_4	4	1306.2	995.6

Table 5.6 Resonant frequencies of CH_3Cl structure molecules (cm^{-1}) from Herzberg (1945).

	Degeneracy	CH_3Cl	CH_3F	CH_3Br	CH_3I	CD_3Cl	CD_3Br
ν_1	1	2966.2	2964.5	1972	2968.8	2161	2151
ν_2	1	1354.9	1475.3	1305.1	1251.5	1029	987
ν_3	1	732.1	1048.2	611	532.8	695	577
ν_4	2	3041.8	2982.2	3055.9	3060.3	2286	2293
ν_5	2	1454.6	1471.1	1445.3	1440.3	1058	1053
ν_6	2	1015.0	1195.5	952.0	880.1	775	717

References

Abragam A and Bleaney B 1970 *Electron Paramagnetic Resonance of Transition Ions* (Oxford: Clarendon)

Banwell C N 1966 *Fundamentals of Molecular Spectroscopy* (New York: McGraw-Hill)

Carlson E H and Dieke G H 1961 *J. Chem. Phys.* **34** 1602

Colthup N B, Daly C H and Wiberly S E 1975 *Introduction to Infrared and Raman Spectroscopy* (New York: Academic)

Cotton F A 1963 *Chemical Applications of Group Theory* (New York: Interscience)

Cowan R D 1981 *The Theory of Atomic Structure and Spectra* (Berkeley: University of California Press)

Crosswhite H M and Dieke G H 1961 *J. Chem. Phys.* **35** 1535

Dieke G H and Pandey B 1964 *J. Chem. Phys.* **41** 1952

Elliott J P and Dawber P G 1979 *Symmetry in Physics* vol. I (London: Macmillan)

Griffith J S 1961 *The Theory of Transition Metal Ions* (Cambridge: Cambridge University Press)

Gruber J B and Satten R A 1963 *J. Chem. Phys.* **39** 1455

Herzberg G 1945 *Molecular Spectra and Molecular Structure* vol. II (Princeton, NJ: Van Nostrand) p 306–15

Hüfner S 1978 *Optical Spectra of Transparent Rare-Earth Compounds* (New York: Academic)

Hutchings M T 1964 *Solid State Phys.* **16** 227 (New York: Academic)

Leech J W and Newman D J 1969 *How to Use Groups* (London: Science Paperbacks)

Newman D J 1981 *J. Phys. A: Math. Gen.* **14** L429

Nussbaum A 1971 *Applied Group Theory for Chemists, Physicists and Engineers* (Englewood Cliffs, NJ: Prentice-Hall)

Scott P D, Meissner H E and Crosswhite H M 1969 *Phys. Lett.* **28A** 489

Stevens K W H 1952 *Proc. Phys. Soc.* A **65** 209

Wilson E B, Decius J C and Cross P C 1980 *Molecular Vibrations* (New York: Dover)

Chapter 6

The Finite Element Method

6.1 Introduction

In general, the theoretical representation of a physical system is not unique, emphasis on different aspects of the interactions in it gives rise to quite different mathematical formalisms. Abstraction from local behaviour, e.g. the balance of forces or the rate of flow locally, leads to the partial differential equations for some function describing the system. An alternative approach is to emphasise that the net effect of all local interactions must satisfy some general principles, e.g. the conservation of energy, or some other *global* principle. In mathematical terms this will give rise to quite a different equation for the function, e.g. the temperature distribution, describing the system. In Chapter 3 we have seen how finite difference methods may be employed in the solution of partial differential equations, not, however, without some complexity in the case of systems with complicated boundary geometry or where normal derivative boundary conditions are specified. In this chapter we describe a technique which largely overcomes difficulties of this kind and which is based on a global view of the problem.

To illustrate the idea behind the method of finite elements we consider the problem of determining the electrostatic potential $\Phi(r)$ in a medium populated by n charged conductors, the potential on each of which, Φ_i, is specified. The interaction energy of all the charges in the system may be written as an integral involving the function Φ; at equilibrium the charges on the conductors would be expected to arrange themselves to maintain each surface at an equipotential and, simultaneously, minimise this energy. From first principles this energy is

$$U = \frac{1}{2} \sum_{i=1}^{n} \int_{S_i} \Phi_i \sigma \, \mathrm{d}S \tag{6.1}$$

where each integral is taken over the surface of the ith conductor, on which the charge density is σ and the potential is Φ_i (a constant). Following standard methods of considering each such surface, together with the sphere at infinity, as bounding the volume exterior to the conductors (see, for example, §1.5.6 of Grant and Phillips (1975)), while noting that the volume charge density everywhere is zero, this energy may be written as

$$U(\Phi) = \frac{\varepsilon_0}{2} \int \varepsilon_r |\text{grad } \Phi|^2 \, \mathrm{d}V \tag{6.2}$$

where ε_r is the relative permittivity of the medium and the integration is over all space exterior to the conductors. It can be readily shown (Chapter 19 of Feynman *et al* (1964)) that the function Φ which minimises the integral U must satisfy the differential equation

$$\nabla \cdot (\varepsilon_r \text{ grad } \Phi) = 0.$$

subject to the specified boundary potentials.

Finding a solution for this differential equation may therefore be viewed in the alternative light of finding a function which will minimise the integral $U(\Phi)$. For many differential equations an equivalent problem of determining the function which will minimise (or maximise) some integral may be formulated, although the specific form of the integral may not be obvious a priori; for the electrostatic problem considered above, when distributed charges of density ρ are also present, the relevant integral is (§19 of Feynman *et al* (1964))

$$I(\Phi) = \frac{\varepsilon_0}{2} \int \left(\varepsilon_r |\text{grad } \Phi|^2 - \frac{2}{\varepsilon_o} \rho \Phi \right) \mathrm{d}V. \tag{6.3}$$

The integral I is a function of the unknown function Φ, i.e. it is a functional, and if we can find the function which minimises this functional we have solved the problem. It is unlikely we would be able to guess the exact analytical form of the function Φ. One approach is to guess a plausible function, $\psi(x, y, z, \boldsymbol{\theta})$, containing a number, N, of unknown parameters θ_i, and evaluate the integral $I(\psi)$. We then use the minimum conditions ($\partial I / \partial \theta_i = 0, i = 1, \ldots, N$) to obtain N equations which may be solved for the parameters θ_i. The solution ψ will still be an approximation to the function defined by the differential equation, just as it was in the finite difference method, but can in practice be a very good approximation, even for a small number of parameters. This is illustrated for a simple case by Feynman in §19 of Feynman *et al* (1964).

The *finite element method* (FEM) is a special case of this approach where the (unknown) values of the function Φ at a number of discrete points (nodes, which define segments) in the region are taken as the parameters and the trial function ψ is taken as a piecewise interpolation polynomial over the region, each piece interpolating over a segment. For example, if the region of the problem were segmented into triangles, and linear interpolation used, the approximation to Φ in any one triangle would be written as (see equation (1.8))

$$\psi(x, y) = \sum_{i=1}^{3} \Phi_i N_i(x, y) \tag{6.4}$$

where the functions $N_i(x, y)$ are given by (1.9). Minimising the integral I will give a set of equations from which the value of the potential at the nodes where it is unknown may be determined. The final solution, as in the finite difference method, thus consists of estimates of the value of the function at these nodes. Because the function used in minimising the functional is an approximation, the interpolating function (6.4), the values obtained at the nodes on carrying out the minimisation are, in general, not the exact values of the function which would minimise the exact functional. Hence, as in the finite difference method, they are denoted by ϕ, i.e. the analogue of a truncation error exists which may be reduced by reducing the size of the segmentation or by using higher order functions as interpolants.

For a general problem the integral, or functional, to be minimised must be determined from some considerations of the type discussed above. In §6.5 we discuss the general aspects of replacing a differential equation by a variational principle and show that for a wide class of important problems the functional is given by (6.8). We also show there that the only boundary conditions which the trial function ψ need satisfy are of the Dirichlet type, i.e. where the value of $\Phi = F$ is explicitly given. At other points on the surface where boundary conditions are of the mixed type they will be *automatically* satisfied by the trial function which minimised the functional. Using piecewise polynomials will, by definition, ensure that nodes which fall on the boundary assume values specified by any Dirichlet condition applicable to that boundary. Beyond this, detailed knowledge of §6.5 is not necessary for use of the method. Our presentation, for reasons of simplicity and ease of comparison with the finite difference method in Chapter 3, will be for potential problems. However, the method is of more general applicability and in a suggested project later in the chapter we indicate how it may be applied to eigenvalue problems in quantum mechanics.

6.2 The Method

There is an important class of differential equations which are of the form

$$-\operatorname{div}(p \operatorname{grad} \Phi) + g\Phi = \rho \tag{6.5}$$

with Dirichlet boundary conditions on at least part of the boundary Γ, that is

$$\Phi = F(s) \tag{6.6}$$

and on the remainder Neumann, or mixed, conditions which we write in the form

$$p(s)\frac{\partial \Phi}{\partial n} + q(s)\Phi = b(s). \tag{6.7}$$

For this class, it is shown in §6.5 that the functional to be minimised is given by

$$I(\psi) = \int_{V(\Gamma)} (p|\text{grad}\,\psi|^2 + g\psi^2 - 2\rho\psi)\,\mathrm{d}V + \int_{S'(\Gamma)} (q\psi^2 - 2b\psi)\,\mathrm{d}S. \quad (6.8)$$

Here, the integrals are evaluated over the volume (area) of the region bounded by Γ and over that part of its surface (perimeter), S', on which the mixed type boundary conditions are specified, of the region bounded by Γ. It is also shown in §6.5 that if a function is found which satisfies condition (6.6) where appropriate, and minimises this functional, it is the solution to the problem. It is not necessary for such a trial function to satisfy relations of the type (6.7), the minimisation of (6.8) ensures that it will *automatically* satisfy these 'natural' boundary conditions. This is a great advantage in solving problems where, as we saw using the finite difference method, boundary conditions involving derivatives can be very troublesome, especially if the boundary does not coincide with the mesh.

To illustrate the method, we consider Poisson's equation in the unit square

$$-\,\text{div}(\varepsilon\ \text{grad}\ \Phi) = \rho \quad (6.9)$$

with the potential given everywhere on the perimeter, where $\Phi = F(s)$. The region is divided into area elements, we choose the simplest case of triangular elements. These define a total of N nodes, where we denote the potential by ϕ_μ, $\mu = 1, \ldots, N$. In figure 6.1 we show an arbitrary segmentation of the region using nine nodes, which define ten triangles. At the vertices of any triangular element (e), we use the duplicate notation $\phi_1^{(e)}, \phi_2^{(e)}, \phi_3^{(e)}$ to relabel its nodes locally and write the trial function ψ in that element as the linear interpolant (6.4), that is

$$\psi^{(e)}(x, y) = \sum_{i=1}^{3} N_i^{(e)}(x, y)\phi_i^{(e)}$$

where $N_i^{(e)}$ are the shape functions for linear interpolation on a triangle, given by (1.9), and $\psi^{(e)} = 0$ if (x, y) does not lie in the element (e). Over the whole region we write

$$\psi = \sum_{(e)} \psi^{(e)}(x, y).$$

If the element (e) has some of the specified values F_i as nodes, this form of the function will satisfy the essential boundary condition since the interpolant reduces to the nodal values when evaluated at its nodes. With our problem and boundary conditions the functional (6.8) reduces to ($p = \varepsilon$,

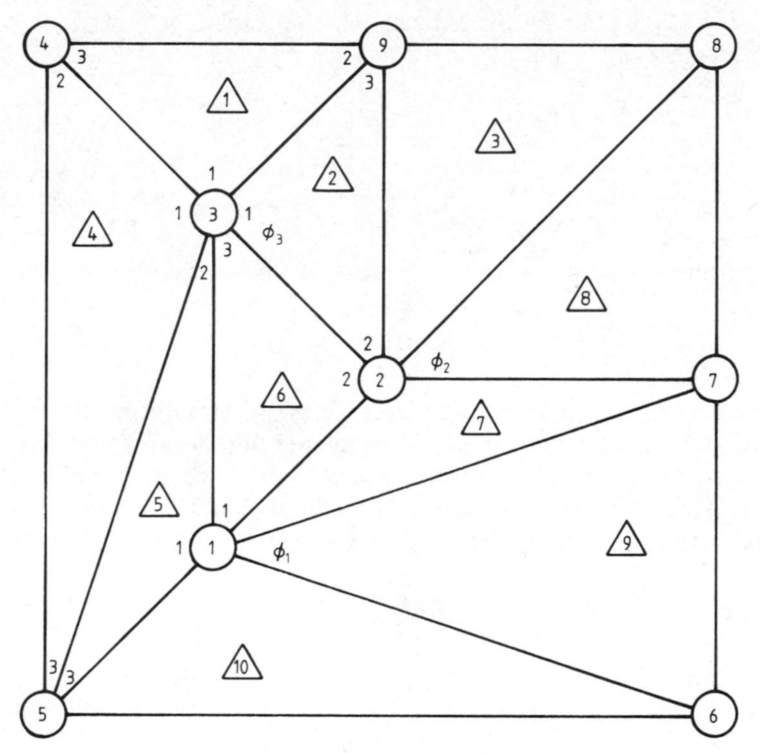

Figure 6.1 An arbitrary segmentation of the square into ten triangular elements, in each of which the trial function $\psi(x, y)$ is approximated by a linear interpolation, using the nodal values of the sought after solution as parameters.

$g = 0$, $S' = 0$)

$$I(\psi) = \int \mathrm{d}x\,\mathrm{d}y \left[\varepsilon\left(\frac{\partial \psi}{\partial x}\right)^2 + \varepsilon\left(\frac{\partial \psi}{\partial y}\right)^2 - 2\rho\psi \right]$$

$$= \sum_{(\mathrm{e})} \int_{\mathrm{e}} \mathrm{d}x\,\mathrm{d}y \left\{ \varepsilon\left[\left(\frac{\partial \psi^{(\mathrm{e})}}{\partial x}\right)^2 + \left(\frac{\partial \psi^{(\mathrm{e})}}{\partial y}\right)^2\right] - 2\rho\psi^{(\mathrm{e})} \right\} \tag{6.10}$$

since the only non-zero contribution to ψ in an element (e) is $\psi^{(\mathrm{e})}$. Because we are using a linear interpolation function the derivatives are constants, for example (cf equation (1.9))

$$\frac{\partial \psi^{(\mathrm{e})}}{\partial x} = \sum_{i=1}^{3} \frac{\partial N_i^{(\mathrm{e})}}{\partial x} \phi_i^{(\mathrm{e})} = \frac{1}{2A^{(\mathrm{e})}} [(y_2 - y_3)\phi_1^{(\mathrm{e})} + (y_3 - y_1)\phi_2^{(\mathrm{e})} + (y_1 - y_2)\phi_3^{(\mathrm{e})}]$$

where $A^{(e)}$ is the area of the element (e). Considering just the contribution from element (e) we have

$$\left[\left(\sum_{i=1}^{3} \frac{\partial N_i^{(e)}}{\partial x} \phi_i^{(e)}\right)^2 + \left(\sum_{i=1}^{3} \frac{\partial N_i^{(e)}}{\partial y} \phi_i^{(e)}\right)^2\right] \times \int_{e} \varepsilon \, \mathrm{d}x \, \mathrm{d}y - \sum_{i=1}^{3} \phi_i^{(e)} \int_{e} 2\rho N_i^{(e)} \, \mathrm{d}x \, \mathrm{d}y.$$

This expression contains several integrals which are problem specific, e.g.

$$\int_{e} \rho(x, y) N_i^{(e)} \, \mathrm{d}x \, \mathrm{d}y.$$

These may of course be carried out analytically, or numerically, in a given problem. However, it is common to encounter library programs in this field which enable general problems to be solved. One can, for example, approximate the function ρ in the element (e), even though it is completely known there, by linear interpolation using its value at the vertices, that is

$$\rho(x, y) = \sum_{j=1}^{3} \rho_j N_j(x, y).$$

This is advantageous since these integrals then take the form

$$\sum_{j=1}^{3} \rho_j \int_{e} N_j^{(e)} N_i^{(e)} \, \mathrm{d}x \, \mathrm{d}y.$$

The basis functions N_i are dimensionless so the integrals occurring here are simply multiples of the area of the element, $A^{(e)}$, and are independent of its position and orientation. General expressions for the values of integrals of products of the basis functions over an element are derived in Appendix 2 of Davies (1980). For our purpose we just need the results, given by

$$\begin{aligned} t_{ii}^{(e)} &\equiv 2 \int_{(e)} N_i^2 \, \mathrm{d}x \, \mathrm{d}y = A^{(e)}/3 \\ t_{ij}^{(e)} &\equiv 2 \int_{(e)} N_i N_j \, \mathrm{d}x \, \mathrm{d}y = A^{(e)}/6 \qquad i \neq j. \end{aligned} \tag{6.11}$$

A similar expansion for the integral containing ε may be used; for simplicity we will assume ε is constant. We can then write the contribution of element (e) to the functional as:

$$I^{(e)} = \sum_{i,j=1}^{3} k_{ij}^{(e)} \phi_i^{(e)} \phi_j^{(e)} - \sum_{i=1}^{3} R_i^{(e)} \phi_i^{(e)}$$

where

$$k_{ii}^{(e)} = \frac{\varepsilon^{(e)}}{4A^{(e)}}[(y_j - y_k)^2 + (x_j - x_k)^2]$$

$$k_{ij}^{(e)} = \frac{\varepsilon^{(e)}}{4A^{(e)}}[(y_j - y_k)(y_k - y_i) + (x_j - x_k)(x_k - x_i)] \qquad i \neq j$$

$$k_{ji}^{(e)} = k_{ij}^{(e)} \tag{6.12}$$

$$A^{(e)} = \tfrac{1}{2}[(x_1y_2 - x_2y_1) + (x_2y_3 - x_3y_2) + (x_3y_1 - x_1y_3)]$$

$$R_i^{(e)} = \sum_{j=1}^{3} t_{ij}^{(e)}\rho_j^{(e)}.$$

We rewrite the sums in terms of the global coordinates, that is

$$I^{(e)} = \sum_{\mu,\nu=1}^{N} k_{\mu\nu}^{(e)}\phi_\mu\phi_\nu - \sum_{\mu=1}^{N} R_\mu^{(e)}\phi_\mu$$

where, of course, the great majority of the coefficients are zero. For the element 2 in the figure, only coefficients with μ and ν having values two, three and nine are non-zero. If we now add up the contribution from all the elements, equation (6.10), we can write

$$I(\psi) = \sum_{\mu=1}^{N}\left(\sum_{\nu=1}^{N} K_{\mu\nu}\phi_\mu\phi_\nu - R_\mu\phi_\mu\right) \tag{6.13}$$

where

$$K_{\mu\nu} = K_{\nu\mu} = \sum_{(e)} k_{\mu\nu}^{(e)} \qquad \text{and} \qquad R_\mu = \sum_{(e)} R_\mu^{(e)}.$$

For example, referring to figure 6.1, for $\mu = \nu = 3$ we have

$$K_{33} = k_{33}^{(1)} + k_{33}^{(2)} + k_{33}^{(4)} + k_{33}^{(5)} + k_{33}^{(6)}.$$

At a number r ($r \geqslant 1$) of the nodes, Φ will be specified by the Dirichlet boundary condition; for convenience we label these points $(N - r + 1), \ldots, N$ and denote the values by F_α. Whatever numbering scheme is adopted, the final equations can always be renumbered to correspond to this scheme.

Minimising $I(\psi)$ with respect to the unknown values of the potential then gives us $(N - r)$ equations, as shown by

$$\frac{\partial I}{\partial \phi_\mu} = \sum_{\nu=1}^{N} 2K_{\mu\nu}\phi_\nu - R_\mu = 0 \qquad \mu = 1, \ldots, (N - r).$$

Separating out the known values of the potential, we can write this as

$$\sum_{\nu=1}^{N-r} K_{\mu\nu}\phi_\nu = \frac{R_\mu}{2} - \sum_{\alpha=N-r+1}^{N} K_{\mu\alpha}F_\alpha$$

that is,

$$\mathbf{K}\boldsymbol{\phi} = \boldsymbol{B} \tag{6.14}$$

where $\mathbf{K}$ is an $(N-r)$ by N matrix and the known vector $\mathbf{B}$ has $(N-r)$ components given by

$$B_\mu = (R_\mu/2) - \sum_{\alpha=N-r+1}^{N} K_{\mu\alpha}F_\alpha. \tag{6.15}$$

The simultaneous equations in (6.14) remain to be solved. In practice, some thought must go into the numbering of the nodes so that the matrix $\mathbf{K}$ may have optimum properties. It is clear that $\mathbf{K}$ is symmetric and will have many zero elements; all pairs μ, ν, where μ and ν are not common to any element in the space, give a zero element $K_{\mu\nu}$, e.g. in figure 6.1 $K_{1\nu} = 0$, $\nu = 4$, 8 or 9, $K_{37} = 0$ etc. By suitable numbering of the nodes, it can be arranged for $\mathbf{K}$ to have a sparse-banded structure and special methods have been devised for solving the resulting simultaneous equations by Gaussian elimination, e.g. see §6.3 of Cheney and Kincaid (1980) and also Bathe (1982). Iteration methods of the type discussed in §3.3 of this book may also be used. In this respect we note that, since the solution at a particular node is determined only by the values at nodes belonging to segments to which this node is also common, a systematic and tidy segmentation to keep the number of nodes belonging to the segments constant will be helpful.

Obvious advantages of the FEM include the flexibility it allows for handling non-simple geometrical boundaries (any curve can be approximated to arbitrary accuracy using linear segments) and the fact that the density of nodes can be chosen to concentrate on important areas where the variation of the function is most rapid.

6.3 An Example

In practical calculations a systematic method of numbering must be adopted and the contributions of the various elements to the functional are best evaluated in terms of simplex coordinates, introduced in the next section. A clear example of the procedures is to be found in the FORTRAN program listing for a simple problem, similar to that being treated here, given in Chapter 1 of Silvester and Ferrari (1983). Without this complication, we illustrate the main components of the calculation for our problem above, (6.9), taking $\varepsilon = 1$, $\rho(x, y) = 2y$ and boundary conditions

$$\Phi(0, y) = 0 \qquad \Phi(x, 0) = 0 \qquad \Phi(1, y) = 0$$

$$\Phi(x, 1) = x(1-x).$$

For each element a block of data must be prepared listing the global and local node numbers, their coordinates and the values of all relevant functions there. For illustration purposes the segmentation in figure 6.1 involves only three nodes where the potential is unknown (to keep the final solution of the simultaneous equations to a trivial exercise), with coordinates $(\frac{1}{4}, \frac{1}{4})$, $(\frac{1}{2}, \frac{1}{2})$ and $(\frac{1}{4}, \frac{3}{4})$ respectively. The data table for element number 2 in this case would have the following form.

μ	i	$x_i^{(2)}$	$y_i^{(2)}$	$\rho_i^{(2)}$	F_μ
3	1	0.25	0.75	1.5	—
2	2	0.50	0.50	1.00	—
9	3	0.50	1.00	2.00	0.25

Using (6.11) and (6.12) the following quantities may be formed, where the subscripts indicate global coordinates,

$$k_{22}^{(2)}=0.5 \qquad k_{29}^{(2)}=0 \qquad k_{32}^{(2)}=-0.5 \qquad k_{33}^{(2)}=1.0 \qquad k_{39}^{(2)}=-0.5$$

$$A^{(2)}=1/16 \qquad R_2^{(2)}=11/192 \qquad \text{and} \qquad R_3^{(2)}=1/16.$$

Note that only quantities where a subscript involves one of the unknown potentials are required in (6.13), so in this case k_{99} and R_9 are not required. When this has been performed for all the elements the overall matrices are assembled using (6.13), resulting in equation (6.14), which in the present case takes the form:

$$\begin{bmatrix} \frac{19}{3} & -2 & -1 \\ -2 & 5 & -1 \\ -1 & -1 & 5 \end{bmatrix} \begin{bmatrix} \phi_1 \\ \phi_2 \\ \phi_3 \end{bmatrix} = \begin{bmatrix} 0.08854 \\ 0.18229 \\ 0.16146 \end{bmatrix} - \begin{bmatrix} 0 & -3 & -\frac{2}{3} & \frac{1}{3} & 0 & 0 \\ 0 & 0 & 0 & -\frac{3}{2} & 0 & -\frac{1}{2} \\ -2 & 0 & 0 & 0 & 0 & -1 \end{bmatrix} \begin{bmatrix} 0 \\ 0 \\ 0 \\ 0 \\ 0 \\ \frac{1}{4} \end{bmatrix}$$

The solutions are $\phi_1=0.068$, $\phi_2=0.112$ and $\phi_3=0.118$. On comparison with the known solution to this problem, $\Phi(x, y)=xy(1-x)$, we have $\Phi_1=0.047$, $\Phi_2=0.125$ and $\Phi_3=0.141$. Thus the error is as large as 30% for node 1; a greater number of nodes and a more regular segmentation would give results much closer to the true solution.

In general, the matrix $\mathbf{K}$ will be of high dimensions with all non-zero values lying along bands parallel to the diagonal; only values $K_{\mu\nu}$, where nodes μ and ν are common to an element in the segmentation, are non-zero. Quite sophisticated, efficient methods for the solution of the resulting equations have been devised, and the task is probably best entrusted to library programmes. However, one such method is described in Exercise 6.3 at the end of this chapter.

PROJECT 6A: ENERGY LEVELS FOR A MOLECULAR POTENTIAL

A potentially useful application of the finite element methods as described in this chapter is to Schrödinger's equation in two dimensions. Three-dimensional problems having a plane of symmetry can be reduced to this form and the application of the method to some such problems is described in Friedman *et al* (1978) and Duff *et al* (1980). The method is particularly advantageous if the potential is only specified numerically at some points; for example, from experiment. In some cases where the potential contains a singularity its effects may be smoothed out in the integrals over elements which occur in the method.

It is common for a problem to be specified with the wavefunction vanishing on some boundary. In such a case, comparing Schrödinger's time-independent equation

$$\nabla^2\psi + \frac{2m}{\hbar^2}(E - V)\psi = 0$$

with the notation in §6.5, we have $p = -1$, $g = (2m/\hbar^2)(E - V)$, $\rho = 0$ and $F = 0$, so that the functional to be minimised is, cf (6.8),

$$I(\psi) = \iint \left[-|\text{grad}\,\psi|^2 + \frac{2m}{\hbar^2}(E - V)\psi^2 \right] \mathrm{d}x\,\mathrm{d}y.$$

Using the same method of segmentation of the region into triangles, adopting linear interpolation functions as above and approximating the potential $V(x, y)$ in each element by its interpolant, i.e. $V^{(e)}(x, y) = \Sigma\, N_i^{(e)}(x, y) V_i^{(e)}$, an equation corresponding to (6.13) is obtained which has the form

$$I(\psi) = -\sum_{\mu,\nu} (H_{\mu\nu} - ES_{\mu\nu})\phi_\mu\phi_\nu.$$

Here

$$H_{\mu\nu} = k_{\mu\nu} + \frac{m}{\hbar^2}\sum_\lambda t_{\mu\nu\lambda}V_\lambda$$

$$S_{\mu\nu} = (m/\hbar^2)t_{\mu\nu}$$

where $k_{\mu\nu}$ etc are assembled from the corresponding quantities for the individual elements, $t_{ij}^{(e)}$ and $k_{ij}^{(e)}$ are defined in (6.11) and (6.12) respectively and

$$t_{ijk}^{(e)} = 2\int_e N_i N_j N_k \,\mathrm{d}x\,\mathrm{d}y$$

may be obtained from Exercise 6.1. Minimising $I(\psi)$ as before will give rise to the generalised eigenvalue problem

$$(\mathbf{H} - E\mathbf{S})\boldsymbol{\phi} = 0 \qquad (6.16)$$

(cf equation (4.13)), where $\boldsymbol{\phi}$ has as many components as there are interior nodes in the segmentation. Note, however, that in terms of solution this equation is quite different from (4.9). Here, the expansion of ψ is in terms of local shape functions $N_i(x, y)$, with the unknown values of the wavefunction ϕ_i as coefficients, whereas the expansion (4.8) is in terms of wavefunctions ϕ_i

of *another* system (individual atoms) with numerical coefficients a_i to be determined. The truncation error associated with the eigenvalues can be shown to be $O(h^{2p})$ where p, here unity, is the order of the interpolating polynomial and h is the characteristic size of an element.

The solution of the generalised eigenvalue problem is still a formidable matter. One method which reformulates the problem in terms of the solutions to two ordinary eigenvalue problems is described in §4.5 (see also §10.2.5 of Bathe (1982) and Exercise 6.3 below). Another important method appropriate for the symmetric-banded matrices which occur here is that of 'subspace iteration', described in §12.3 of Bathe (1982), where some explicit FORTRAN programs for the task are also given.

A popular model for molecular structure is to assume that the molecule is composed of individual oscillators coupled by some anharmonic interaction. An interesting potential related to the stability of planar molecules is the Henon–Heiles potential

$$V(x, y) = \tfrac{1}{2}\mu\omega^2(x^2 + y^2) + \lambda y(x^2 - \tfrac{1}{3}y^2)$$

where μ is the mass of the particle.

Taking $(\hbar/\mu\omega)^{1/2}$ as the unit of length and writing

$$\lambda' = \frac{\lambda}{\hbar\omega}\left(\frac{\hbar}{\mu\omega}\right)^{3/2}$$

Schrödinger's equation assumes the form

$$\nabla^2\psi + 2(E' - V')\psi = 0$$

where $E' = E/\hbar\omega$ (equivalent to taking $\mu = \hbar = 1$) and

$$V'(x, y) = \tfrac{1}{2}(x^2 + y^2) + \lambda' y(x^2 - \tfrac{1}{3}y^2).$$

The finite element method, as described above, should be used to find the eigenvalues of this equation with $\lambda' = 1/(4\sqrt{5})$. Strictly speaking, the boundary condition is $\psi \to 0$ at infinity, in the units used here it will be good enough to take $x, y \gg 1$ on the boundary. A square of side 10 is suggested as the region. The values obtained for E' may be compared with those tabulated in table 1 of Noid and Marcus (1979).

Unlike the illustrative example in §6.3 above, an orderly segmentation with a systematic numbering scheme for the nodes should be used. We note that if d is the largest difference between node numbers occurring in any one element, the total bandwidth of the resulting matrices will be $2d + 1$. By bandwidth we mean the maximum span between non-zero values in any one row of the matrix; in general, the time required to solve the equations will be proportional to this bandwidth. A possible, very rough, segmentation and numbering is shown in figure 6.2(*a*). Here, $d = 5$ with overall bandwidth of 11, all the non-zero elements of **H** and **S** in (6.16) will then lie on 7 diagonals,

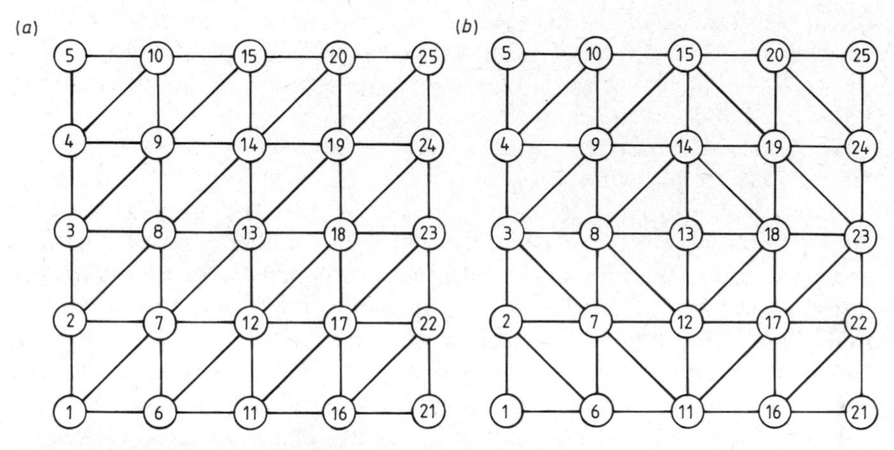

Figure 6.2 Two possible segmentations of the square in Project 6A. In (*a*) the bandwidth of the matrix will be 11, while in (*b*) it will be 13.

although, because of the symmetry of these matrices, only elements on 4 of these diagonals need to be stored. An alternative segmentation, using the same number of elements but with $d = 6$, is shown in figure 6.2(*b*); this has been reported to give more accurate results when a linear interpolant is used, as here. This aspect could be investigated later in the project.

In practice, the elements of **H** and **S** in (6.16) will be constructed from the contributions of individual elements, $H_{ij}^{(e)}$ and $S_{ij}^{(e)}$, where

$$H_{ij}^{(e)} = k_{ij}^{(e)} + \sum_{k=1}^{3} t_{ijk}^{(e)} V'_k$$

$$t_{ijk}^{(e)} = 2 \int_{e} N_i N_j N_k \, dx \, dy$$

$$S_{ij}^{(e)} = t_{ij}^{(e)}$$

with $k_{ij}^{(e)}$, $t_{ij}^{(e)}$ and $t_{ijk}^{(e)}$ defined by (6.12), (6.11) and Exercise (6.1) respectively.

*6.4 Refinements to the Method

So far in this chapter the basic principles of the method of finite elements have been illustrated. In this section we briefly introduce a few ideas which facilitate more accurate application of the method, namely higher order elements, simplex coordinates and isoparametric transformations. More details on these topics may be found in standard texts on the subject; for example, Zienkiewicz (1977).

Improved accuracy of the method can be achieved either by increasing the number of elements or, at the expense of increased complexity, by replacing the linear interpolating functions by higher order polynomials, e.g. a quadratic in two dimensions: $P(x, y) = a_0 + a_1x + a_2y + a_3x^2 + a_4xy + a_5y^2$.

This becomes very involved, six nodal values will be required for the determination of the coefficients and a 6×6 determinant will have to be expanded. To represent these interpolants it is more convenient to introduce local or simplex coordinates and label a point (x, y) in a triangular element by (L_1, L_2, L_3) where, referring to figure 6.3,

$$L_1 = \frac{h}{h_1} = \frac{\text{area BCD}}{\text{area ABC}}$$

$$L_2 = \frac{\text{area CDA}}{\text{area ABC}} \qquad \text{etc.}$$

The relation between the two sets of coordinates $L_i\,(x, y)$ is just the relation (1.9), that is

$$L_i(x, y) = \frac{1}{2A}\left[(x_jy_k - x_ky_j) + (y_j - y_k)x - (x_j - x_k)y\right] \qquad i = 1, 2, 3.$$

The three coordinates are not independent, as is obvious from their definition $L_1 + L_2 + L_3 = 1$, a relation which must be allowed for when differentiating with respect to these coordinates. In terms of these coordinates the shape functions for the linear interpolant are simply

$$N_i(L_1, L_2, L_3) = L_i.$$

Adding three more nodes (4, 5 and 6) on the mid-points of the triangle, we

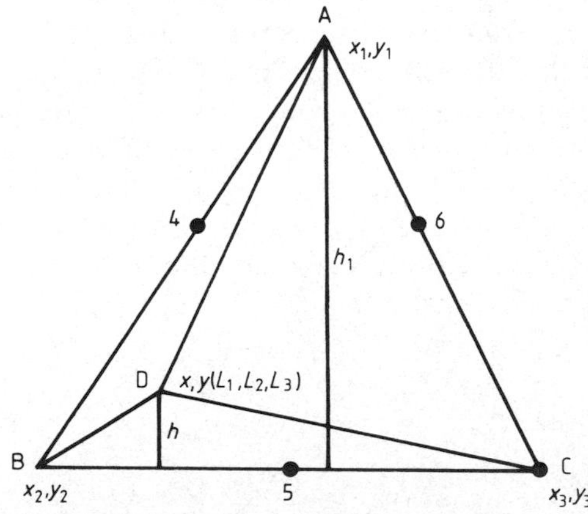

Figure 6.3 Illustrating the relation between the simplex coordinates of a point (L_1, L_2, L_3) and the Cartesian coordinates (x, y). Points numbered 4, 5 and 6 indicate possible locations for the additional nodes required in the element if interpolation using a quadratic is used.

can find the shape functions for the quadratic interpolant, (§5.3 of Davies 1980)

$$\psi(L_1, L_2, L_3) = \sum_{i=1}^{6} N_i(L_1, L_2, L_3)\phi_i$$

$$N_i(L_1, L_2, L_3) = L_i(2L_i - 1) \qquad i = 1, 2, 3$$

$$= 4L_1L_2, 4L_2L_3, 4L_3L_1 \qquad i = 4, 5, 6 \text{ respectively.}$$

Using these local coordinates and shape functions the integrals required to build up the matrix **K** can be expressed in a way analogous to those used to write equation (6.13). In practice, these coordinates would be used for such elements because of the simplicity of the shape elements when expressed in terms of them. However, for purposes of illustration we will continue with the Cartesian coordinates.

The advantage of using higher order interpolants is that larger elements can be used; this may conflict, however, with the requirement of accommodating accurately a curved boundary, which is especially important if the Dirichlet condition is specified there. We now consider a method for taking curved boundaries into account.

The isoparametric transformation is a technique for transforming the variables used as coordinates in an element, e.g. in an individual element integral in (6.10). This transformation has the effect of deforming the element shape so that it comes closer to coinciding with the boundary. In figure 6.4 we show a quadratic triangular element fitted near a curved boundary. In evaluating the contribution to the functional, e.g. leading up to (6.10), it is the area bounded by the curve rather than by its approximation (the triangle) which should be used. We seek a mapping which will map the sides AB and BC as close as possible to the corresponding curved segments. By putting nodes at x_2, y_2 and x_4, y_4, rather than on the sides of the triangle, we can approximate these curves by quadratics. If we denote local coordinates in the triangular element by (ξ, η), we can write the coordinates in the deformed element (x, y) as functions of these, i.e. $x = x(\xi, \eta)$, $y = y(\xi, \eta)$. For example, we require $x(\xi_1, \eta_1) = x_1$, $y(\xi_2, \eta_2) = y_2$, $x(\xi_4, \eta_4) = x_4$ etc. For other points in or on the sides of the triangle we can use our quadratic interpolating functions to interpolate values of x, y, that is

$$x(\xi, \eta) = \sum_{i=1}^{6} N_i(\xi, \eta)x_i$$

$$y(\xi, \eta) = \sum_{i=1}^{6} N_i(\xi, \eta)y_i.$$

This will map the side of the triangle AB on to a quadratic (parabola) passing through (x_1, y_1), (x_2, y_2) and (x_3, y_3), and similarly for side BC, and these curves will be better approximations to the true boundary. We can then

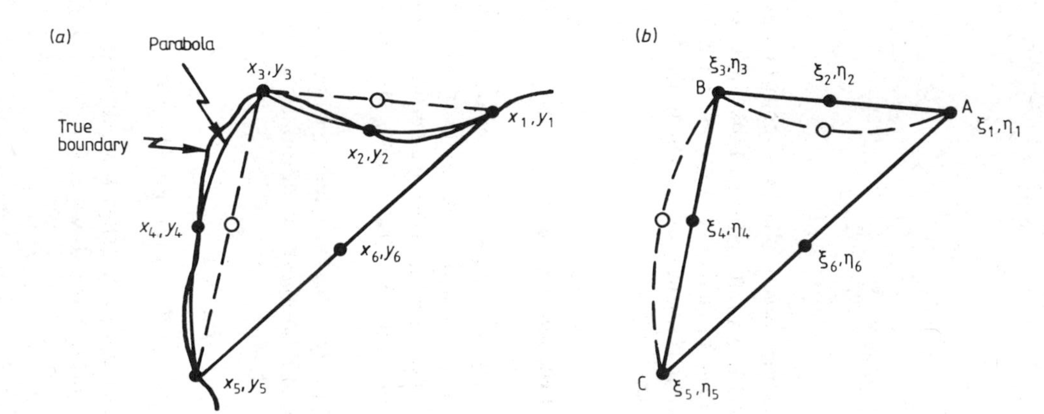

Figure 6.4 (*a*) An irregular boundary is replaced by parts of two quadratics. Coordinates (x_2, y_2) and (x_4, y_4) are assigned to nodes on these curves by mapping nodes, with local coordinates (ξ_2, η_2) and (ξ_4, η_4), originally at the mid-points of the triangular element, as in (*b*).

write integrals over the deformed element (e_T) in terms of integrals over the triangle (e). The contribution for example to (6.10) may be written as

$$I^{(e)}(\psi) = \iint_{e_T} \mathrm{d}x\,\mathrm{d}y \left\{ \dots \right\} = \int_{e} \mathrm{d}\xi\,\mathrm{d}\eta\, J\left(\frac{x,y}{\xi,n}\right) \left\{ \varepsilon \left[\left(\frac{\partial \psi^{(e)}}{\partial \xi}\frac{\partial \xi}{\partial x} + \frac{\partial \psi^{(e)}}{\partial \eta}\frac{\partial \eta}{\partial x} \right)^2 + \left(\frac{\partial \psi^{(e)}}{\partial \xi}\frac{\partial \xi}{\partial y} + \frac{\partial \psi^{(e)}}{\partial \eta}\frac{\partial \eta}{\partial y} \right)^2 \right] - 2\rho\psi^{(e)} \right\}$$

where

$$J \equiv \begin{vmatrix} \dfrac{\partial x}{\partial \xi} & \dfrac{\partial y}{\partial \xi} \\ \dfrac{\partial x}{\partial \eta} & \dfrac{\partial y}{\partial \eta} \end{vmatrix}$$

is the Jacobian of the transformation. The local coordinates, in practice, might be the simplex coordinates L_i introduced above. We will not go into the details of explicit evaluation of elements of the matrix $\mathbf{K}$, nor of obvious generalisations to higher order elements, all of which are discussed in texts on this topic, e.g. Zienkiewicz (1977).

*6.5 Basis in a Variational Principle

In this section we present some arguments for the form of the functional which should be constructed for a wide class of problems. We show that boundary conditions of the non-Dirichlet type are automatically satisfied by the function which minimises this functional and satisfies Dirichlet-type boundary conditions.

Suppose we have a boundary value problem specified by a differential operator L with homogeneous boundary conditions, i.e. $\gamma(s)$ in equation (3.6) is zero, then

$$L\Phi = \rho \tag{6.17}$$

$$\alpha(s)\,\partial\Phi/\partial n + \beta(s)\Phi = 0 \text{ on boundary } \Gamma. \tag{6.18}$$

Let ψ be any other function which satisfies the boundary conditions (in general it would not be difficult to write down such a function) and consider the following *functional*

$$I(\psi) \equiv \int_{V(\Gamma)} (\psi L\psi - 2\psi\rho)\,\mathrm{d}V \tag{6.19}$$

where the integration is carried out over the volume enclosed by the boundary Γ. If L is a *self-adjoint* (Hermitian) *positive definite* operator, this integral has its minimum value when evaluated using $\psi = \Phi$ where Φ is the solution of (6.17). In other words, if we can find the function which minimises the functional (6.19), we have the solution of the differential equation (6.17). To prove this we use the inner product notation, that is

$$(L\psi, \Phi) \equiv \int \Phi L\psi \, \mathrm{d}V.$$

Then L is positive definite if, for *any* function ψ satisfying (6.18), $(L\psi, \psi) > 0$ if $\psi \neq 0$ and is self-adjoint if for any two functions, Φ, ψ satisfying (6.18) $(L\Phi, \psi) = (L\psi, \Phi)$. With this notation (6.19) becomes

$$\begin{aligned} I(\psi) &= (L\psi, \psi) - 2(\rho, \psi) \\ &= (L\psi, \psi) - 2(L\Phi, \psi) \qquad \text{(using 6.17)} \\ &= (L(\psi - \Phi), \psi) - (L\psi, \Phi) \quad \text{(rearranging and using self-adjointness)} \\ &= (L(\psi - \Phi), (\psi - \Phi)) + (L(\psi - \Phi), \Phi) - (L\psi, \Phi) \\ &= (L(\psi - \Phi), (\psi - \Phi)) - (L\Phi, \Phi). \end{aligned}$$

The second term is independent of ψ, and, since L is positive definite, the first term is always positive and has its least value, zero, when $\psi = \Phi$. Hence, the integral I assumes its least value when evaluated with the solution of the problem (6.17). We consider the finite element method for solving (6.17) as a method for solving the alternative but equivalent problem of finding the function Φ which minimises (6.19). In fact the method is much more general.

When the boundary conditions are not homogeneous, i.e. $\gamma(s)$ in (3.6) is not zero, they must also be allowed for in the form of the functional. We will consider the case of the important operator (3.7), $L = -\operatorname{div} p \operatorname{grad} + g$ with general boundary conditions (3.6), which we rewrite as

$$\Phi = F(s) \qquad (6.20a),$$

if $\alpha = 0$, i.e. Φ is specified. Otherwise, for $\alpha \neq 0$, we can write

$$p(s)(\partial\Phi/\partial\hat{n}) + q(s)\Phi = b(s) \qquad (6.20b)$$

where $q = \beta p/\alpha$, $b = \gamma p/\alpha$. If Φ is the solution of $L\Phi = \rho$ and f any function which satisfies the boundary conditions (6.20), then the function $v = \Phi - f$ satisfies

$$p(s)(\partial v/\partial n) + q(s)v = 0$$

and the differential equation is

$$Lv = L\Phi - Lf = \rho - Lf = G(\text{say}).$$

The function v thus satisfies $Lv = G$, subject to homogeneous boundary conditions, and hence $I(v) = I(\Phi - f)$ will be a minimum where I has the form given by (6.19). We can thus solve our problem by seeking the function ψ which satisfies (6.20) and minimises $I(\psi - f)$, which we will henceforth denote simply as $I(\psi)$;

$$\begin{aligned} I(\psi) &= (L(\psi - f), (\psi - f)) - 2(G, (\psi - f)) \\ &= (L\psi, \psi) - (L\psi, f) - (Lf, \psi) + (Lf, f) - 2(G, \psi) + 2(G, f). \end{aligned}$$

The fourth and last terms do not involve ψ at all, so in future minimisation with respect to this function they need not be carried along. Therefore, remembering $G = (\rho - Lf)$, we have

$$
\begin{aligned}
I(\psi) &\to (L\psi, \psi) - 2(\rho, \psi) + (Lf, \psi) - (L\psi, f) \\
&= \int [-\psi \operatorname{div}(p \operatorname{grad} \psi) + g\psi^2 - 2\rho\psi - \psi \operatorname{div}(p \operatorname{grad} f) \\
&\quad + f \operatorname{div}(p \operatorname{grad} \psi)] \, \mathrm{d}V. \qquad (6.21)
\end{aligned}
$$

Using the result $\operatorname{div} f\boldsymbol{A} \equiv f \operatorname{div} \boldsymbol{A} + \boldsymbol{A} \cdot \operatorname{grad} f$, the divergence theorem, and remembering that both ψ and f satisfy (6.20), this reduces to

$$I(\psi) \to \int_{V(\Gamma)} (p|\operatorname{grad} \psi|^2 + g\psi^2 - 2\rho\psi) \, \mathrm{d}V + \int_{S(\Gamma)} (q\psi^2 - 2b\psi) \, \mathrm{d}S \qquad (6.22)$$

where the surface integral is only over those parts of the surface where Φ is not given explicitly and a term $\int bf \, \mathrm{d}s$, which does not contain ψ, has been dropped.

We have shown above that the function ψ which satisfies (6.20) everywhere on the boundary and minimises I is the solution Φ of the differential equation $-\operatorname{div}(p \operatorname{grad} \Phi) + g\Phi = \rho$. If we relax the condition (6.20b) on ψ, i.e. only require ψ to satisfy the boundary condition on those parts of the boundary where it is of the Dirichlet type $\Phi = F(s)$, then if this ψ minimises I the solution sought is still Φ, and the non-Dirichlet boundary conditions are *automatically* satisfied. We now show this. We write $\psi = \Phi + \eta$, where η is an infinitesimal but arbitrary function of position except that it must vanish at those parts of the boundary where the potential is specified, i.e. $\psi = F$ at those parts. When I is a minimum $\delta I = 0$ for such arbitrary functions

$$\delta I = I(\psi) - I(\Phi) = 0$$

or

$$\int (2p \operatorname{grad} \Phi \cdot \operatorname{grad} \eta + 2\eta g\Phi - 2\rho\eta) \, \mathrm{d}V + \int (2q\Phi\eta - 2b\eta) \, \mathrm{d}S = 0.$$

However,

$$p \operatorname{grad} \Phi \cdot \operatorname{grad} \eta = \operatorname{div}(\eta p \operatorname{grad} \Phi) - \eta \operatorname{div}(p \operatorname{grad} \Phi).$$

Thus, rearranging terms and using the divergence theorem once, we find

$$\int [-\operatorname{div}(p \operatorname{grad} \Phi) + g\Phi - \rho]\eta \, \mathrm{d}V + \int \left(p \frac{\partial \Phi}{\partial n} + q\Phi - b\right)\eta \, \mathrm{d}S = 0.$$

The first integral vanishes for arbitrary η because Φ satisfies the differential equation, hence the second integral must also vanish. Its only contributions come from those parts of the surface where the boundary conditions are of the non-Dirichlet type, since elsewhere $\eta = 0$. Thus, if the integral is to vanish, for arbitrary η on these parts the equation

$$p\frac{\partial \Phi}{\partial n} + q\Phi - b = 0$$

must be satisfied. This is just (6.20b). So we see that, provided we impose the Dirichlet boundary conditions on our test function ψ, when it is chosen to minimise I it will automatically satisfy the remainder of the boundary conditions.

In the derivation of the functional (6.22) it has been assumed that the function ψ satisfies the boundary conditions of equation (6.20) everywhere on the boundary. For at least one point on the boundary this condition must be of the Dirichlet type, i.e. Φ must be expressly given somewhere, otherwise the solution would be indeterminate to within a constant. The condition (6.20a) is known as an *essential boundary condition.* The remarkable fact is that, provided ψ satisfies (6.20a) at all regions of the boundary where it holds, the function obtained from minimising (6.22) will automatically satisfy the condition (6.20b) at all other regions of the boundary, i.e. the condition (6.20b) need not be imposed on ψ a priori in these regions. The mixed boundary conditions (6.20b) are *natural boundary conditions* for the functional (6.22), provided the trial function ψ satisfies the Dirichlet boundary condition (6.20a) where applicable.

Exercises

6.1 Relations (6.11) are particular cases of the more general formula

$$\frac{1}{2A}\int N_1^i N_2^j N_3^k \, \mathrm{d}x \, \mathrm{d}y = \frac{i!j!k!}{(i+j+k+2)!}$$

Verify this. (See Appendix 2 of Davies 1980.)

6.2 Show that the expression $\frac{1}{2}\int \Phi\rho \, \mathrm{d}V$ for the energy of the system described in §6.1 reduces to

$$U \to \tfrac{1}{4}\left(\sum_{\mu} R_\mu \phi_\mu + \sum R_\alpha F_\alpha\right)$$

where

$$R_\mu = \sum_{(\mathrm{e})} R_\mu^{(\mathrm{e})}$$

and $R_\mu^{(\mathrm{e})}$ is given in (6.12). Calculate this using the solution obtained in §6.3 and compare it with the known value, 1/18 in this case.

6.3 One method of solving the equation $\mathbf{K}\boldsymbol{\phi} = \boldsymbol{B}$ when $\mathbf{K}$ is symmetric, as in (6.14) above, is to decompose $\mathbf{K}$ into a product of upper and lower triangular matrices, $\mathbf{K} = \mathbf{LU}$ where $L_{ij} = 0$ when $i < j$ and $U_{ij} = 0$ for $i > j$. Then the equation corresponds to the solution of $\mathbf{U}\boldsymbol{\phi} = \boldsymbol{X}$, where the vector $\boldsymbol{X}$ is obtained by solving another equation, $\mathbf{L}\boldsymbol{X} = \boldsymbol{B}$. Since both $\mathbf{U}$ and $\mathbf{L}$ are triangular, both of these equations can be very easily solved. The elements of $\mathbf{U}$ and $\mathbf{L}$ may be obtained as follows: $L_{ij} = \mathrm{sgn}(U_{ii})U_{ji}$; $U_{ij} = S_{ij}$ if S_{ij} is real, $U_{ij} = -|S_{ij}|$ if S_{ij} is imaginary, where S_{ij} is constructed by recursion as described below.

$$S_{11} = K_{11}^{1/2}$$

$$S_{1j} = K_{1j}/S_{11}$$

$$S_{ii} = \left(K_{ii} - \sum_{k=1}^{i=1} S_{ki}^2\right)^{1/2}$$

$$S_{ij} = \left(K_{ij} - \sum_{k=1}^{i=1} S_{ki}S_{kj}\right)\Big/ S_{ii} \qquad \text{for } i < j$$

$$= 0 \text{ for } i > j.$$

As an exercise, use this method to solve the equations at the end of §6.3. This method is known as Cholesky's method, it is discussed in §8.2 of Bathe (1982).

References

Bathe K-J 1982 *Finite Element Procedures in Engineering Analysis* (Englewood Cliffs, NJ: Prentice-Hall)

Cheney W and Kincaid D 1980 *Numerical Mathematics and Computing* (Monterey, CA: Brooks-Cole)

Davies A J 1980 *The Finite Element Method* (Oxford: Clarendon)

Duff M, Rabitz H, Askar A, Cakmak A and Ablowitz M 1980 *J. Chem. Phys.* **72** 1543–59

Feynman R P, Leighton R B and Sands M 1964 *The Feynman Lectures on Physics* vol. II (Reading, MA: Addison-Wesley)

Friedman M, Rosenfeld Y, Rabinovitch A and Thieberger R 1978 *J. Comput. Phys.* **26** 169–80

Grant I S and Phillips W R 1975 *Electromagnetism* (London: John Wiley)

Noid D W and Marcus R A 1979 *J. Chem. Phys.* **67** 559–67

Silvester P P and Ferrari R L 1983 *Finite Elements for Electrical Engineers* (Cambridge: Cambridge University Press).

Zienkiewicz O C 1977 *The Finite Element Method* 3rd edn (London: McGraw-Hill)

Chapter 7

Monte Carlo Methods

7.1 Introduction

The Monte Carlo technique is a numerical method for obtaining an estimate of the solution of a specified problem by using a sequence of values of a random variable. These variables may be used either in a mathematical simulation of the physical behaviour underlying the problem, or as a replacement of the original problem by a stochastic problem which has the same solution. Its colourful name, originally that of a secret file on the Manhattan project, arose from the fame of the casino at Monte Carlo which uses the same fundamental principles in its operation. It has found wide application in both the physical and biological sciences. Within physics we may cite its importance in nuclear physics, e.g. in calculating the extent to which particles penetrate a shielding medium or the related phenomenon of the development of a cascade produced by a primary cosmic ray in the atmosphere. Another widespread application is in statistical mechanics where it provides approximate solutions for otherwise intractable problems in the properties of polymers and on the thermodynamic behaviour of interacting multiparticle systems, notably the occurrence, or otherwise, of phase transitions in such systems. This latter application can be generalised to interacting fields and applied to predict expected states of quark–gluon interactions in high energy physics. Examples of the application of Monte Carlo methods in these fields are given in Chapter 8. Finally, we should mention the role played by Monte Carlo simulations in predicting the performance of complex experiments and in analysing the data from such experiments.

The problems to which Monte Carlo methods can be applied can be distinguished as being of two types, random or deterministic. A random physical process is a phenomenon where the dependence of the probability of its occurring in any time interval $\mathrm{d}t$ is simply $\mu\,\mathrm{d}t$, where μ is a constant. It does not depend on the actual time t relative to any other event. The coefficient μ, which represents the 'strength' of the probability, is a quantity characteristic of the process. Starting from any arbitrary origin of time, we denote by $P(t)$ the probability that the process will not have occurred by the time t. Then, since $(1 - \mu\,\mathrm{d}t)$ is the probability that it will not occur in the succeeding interval $\mathrm{d}t$, we have by definition

$$P(t + \mathrm{d}t) = P(t)(1 - \mu\, \mathrm{d}t).$$

From this, a simple differential equation follows which has the solution

$$P(t) = \mathrm{e}^{-\mu t}.$$

Two questions immediately suggest themselves. What is the probability that two successive occurrences of the event are separated by a time interval in the range $(t, t + \mathrm{d}t)$? Also, what is the probability of exactly n events occurring in a given interval of time? (Of course, the case $n = 0$ is just $P(t)$.) The probability $p(t)\,\mathrm{d}t$ that the time interval between two events lies in the range $(t, t + \mathrm{d}t)$ is just the probability of it not occurring by time t multiplied by the probability that it will occur in $\mathrm{d}t$, i.e.

$$p(t)\,\mathrm{d}t = P(t)\mu\,\mathrm{d}t = \mu\mathrm{e}^{-\mu t}\,\mathrm{d}t.$$

Thus the time separation of events is given by the exponential distribution (A.9) in §A.2 of the Appendix and the mean separation between events is

$$\tau \equiv \int tp(t)\,\mathrm{d}t = \mu^{-1}$$

i.e. $\mu = 1/\tau$. The average number of events occurring in an interval t is just μt. The probability of getting exactly n events inside an interval t can be shown (Braddick 1966) to be given by the Poisson distribution (A.15), i.e.

$$p(n) = \frac{(\bar{n})^n\,\mathrm{e}^{-\bar{n}}}{n!}$$

where $\bar{n} = \mu t$; μ of course will depend on physical parameters. In the case of the decay of a particle, μ depends, because of time dilation, on its velocity; for shot noise in a thermionic tube μ depends on the temperature of the filament, and for the interaction of a proton in a bubble chamber it depends on the particle's energy, possibly spin, magnetic moment etc.

The Monte Carlo method is a natural approach in problems of this kind which contain a stochastic element, i.e. where inherently random behaviour such as Brownian motion, radioactive decay of a particle or the location of fission by a neutron in a fissile medium plays a role. It is also widely used in problems in which no non-deterministic component is obvious, e.g. in the evaluation of multidimensional phase space integrals, the prediction of the specific heat of a polymer material, or the solution of a partial differential equation. A brief consideration of the latter topic, with which we are quite familiar from Chapter 3, will serve as a convenient introduction to the more important aspects of the technique, and in particular serve to introduce the concept of a random walk. Among other important concepts is the idea that our solution can never be unique but is always an estimate, and thus has a variance or standard deviation uncertainty associated with it. The determination of this uncertainty, its reduction, and its significance are vital aspects of Monte Carlo calculations. The statistical concepts required are summarised in §§A.1–A.3 of the Appendix.

7.2 A Random Walk

It is assumed that an unlimited sequence of random numbers, ξ_i, is available from a uniform distribution $0 \leqslant \xi_i \leqslant 1$ (the generation of such numbers will be considered in §7.3). We will consider their use in solving the difference equation arising from Poisson's equation, $\nabla^2\Phi = q$ with Φ given on the boundary Γ by a specified function $F(s)$, where s is a coordinate on Γ. The resulting difference equation on a rectangular grid, i.e. equation (3.9) with $f = 0$, can be rewritten to express the value of ϕ at the point 0 (cf figure 3.1) in terms of its value at the four nearest-neighbour points as in (3.12), that is

$$\phi_0 = \sum_{j=1}^{4} W_{0j}\phi_j + \mu q_0 \tag{7.1}$$

where

$$\begin{aligned} W_{01} &= W_{03} = \tfrac{1}{2}(\lambda^2 + 1)^{-1} \\ W_{02} &= W_{04} = \lambda^2 W_{01} \\ \mu &= -h^2 W_{01}. \end{aligned} \tag{7.2}$$

The term q_0 is the value of q at the point 0 and $\lambda = h/k$. We notice that $\Sigma_{j=1}^{4} W_{0j} = 1$ allowing us to interpret the W_{0j} values as probabilities. If we choose one of the four points, say $j = m$ with an a priori probability W_{0m}, we would have an estimate of ϕ_0, which we denote by $z^{(1)}$, by taking just the value of ϕ at that point

$$z^{(1)} = \phi_m + \mu q_0. \tag{7.3}$$

We would make this decision on the basis of a single random number ξ, e.g. if $\xi < W_{01}$ we choose point 1, if $\xi < (W_{01} + W_{02})$ we choose point 2 etc. Note that, in this case, if $h = k$ all the probabilities are equal to 1/4. In general however they will be a function of position. Expression (7.3) is an estimate in the sense that if the process were repeated many times, every time choosing one of the adjacent points m with probability W_{0m}, the average of the estimates of $\phi(x_0, y_0)$ obtained would tend to the value given by (7.1). The problem with a single estimate, (7.3), is that in general (i.e. unless the point m lies on the boundary Γ) we do not know the potential ϕ_m. We can, however, estimate ϕ_m in a similar way using (7.1), i.e. with a probability W_{mn} select ϕ_n at one of the four points surrounding the point m, so as an estimate of ϕ_m we have $(\phi_n + \mu q_m)$. Inserting this in (7.3) gives us a new estimate of ϕ_0,

$$z^{(2)} = \phi_n + \mu(q_0 + q_m).$$

The quantity $z^{(2)}$ is known as the score of the random walk after two steps. The general method should be clear; we execute a random walk from the point at which we wish to determine the potential, proceeding from a point m to the next point n at random on the basis of the *transition probabilities* W_{mn}, and accumulating the score, until a point on the boundary is reached, in which case no further estimation is required. In a case as in figure 3.4, some decision must be made on striking a boundary where $\partial\Phi/\partial n$, rather than Φ, is specified, e.g. at point s if the next random decision were to favour going outside to pont t. In the case where $\partial\Phi/\partial n = 0$ it is simple to argue replacing the step by a step backwards to u, i.e. the transition probabilities are modified. In the more general case some modification to the score would be necessary.

If j steps are required before the boundary is reached, our final estimator of ϕ_0 is the score of the walk given by

$$z^{(j)} = F_j + \sum_{k=0}^{j-1} \mu q_k \tag{7.4}$$

where the summation runs over all the lattice points encountered in the walk, including the starting point, but not the final point on the boundary. Since such a walk may end up anywhere on the boundary, it is not likely to give a reliable estimate of ϕ. By taking a series of N such random walks all starting at the point x_0, y_0 we obtain a series of estimates, z_i $(i = 1, \ldots, N)$, and from them a mean $\bar{\phi}_0 \pm \sigma_{\bar{\phi}_0}$. The first and second terms are given by

$$\bar{\phi}_0 = \sum_{i=1}^{N} z_i/N$$

and

$$\sigma_{\bar{\phi}_0} = \sigma_{\phi_0}/\sqrt{N}$$

respectively. The variance of the series, $\sigma^2_{\phi_0}$, is calculated using

$$\sigma^2_{\phi_0} = \frac{N}{N-1}\left[\langle z^2\rangle - (\bar{\phi}_0)^2\right].$$

The method can be readily extended to more general elliptic equations in two and three dimensions but it is more useful as an illustration of elementary Monte Carlo methods and the principles of estimation generally, than as a practical tool. It may nevertheless be a viable alternative to the other methods described in this book for solving such equations in cases where the value of the function is only required at one point in a system with complicated geometry.

A random walk or chain of the type just described, where the transition probabilities at any stage are not constrained by the previous history of the walk (e.g, sites can be revisited any number of times) is known as a Markov process, and because it must eventually strike the boundary the walk described is called an absorbing Markov chain. We will encounter later on, in §8.6, an example of a non-Markovian process, a *self-avoiding* random walk where the transition probabilities at any stage have to take into account the previous history of the walk, and in a case like the above the transition probabilities may result in a forced termination of the walk before the boundary is ever reached. Random walks are also useful in a more abstract context, e.g. in considering the possible members of an ensemble of statistical mechanical systems. We shall see how representative states characteristic of a system in equilibrium, i.e. obeying the Boltzmann distribution, can be sampled using such a notional random walk on the system in §7.4

The above example of the use of a random walk for solving a difference equation illustrates the principal features of any Monte Carlo calculation, notably that the answer, or score, obtained as a function of a sequence of random numbers, is an estimate of the solution sought. Associated with such an estimate is the variance of its distribution. The smaller this variance, the less the uncertainty in the determination, so that, although basically a very simple method, refinements aimed at reducing the variance for a given amount of computation can result in considerable sophistication. The task of variance reduction can be viewed as being equivalent to finding an algorithm for estimating a multidimensional integral using a fixed number of random variables such that it will have a minimum uncertainty. Any Monte Carlo calculation can be viewed as the evaluation of such an integral. By the mean value theorem, we know that

$$I \equiv \int_a^b f(x)\,\mathrm{d}x = (b-a)f(\zeta)$$

for some ζ in the range $a < \zeta < b$. Hence, an unbiased estimate of I is $(b-a)f(\zeta)$ if ζ is a value for x chosen uniformly at random on (a, b). In the same way, a function of n such variables, $f(\xi_1, \xi_2, \ldots, \xi_N)$, where the ξ_i are selected at random from a uniform distribution on (0, 1), can be considered an estimate of the N-dimensional integral

$$I = \int_0^1 \mathrm{d}x_1 \int_0^1 \mathrm{d}x_2 \ldots\ldots \int_0^1 \mathrm{d}x_N f(x_1, x_2 \ldots . x_N). \qquad (7.5)$$

However, we shall find that two algorithms using the same set of values $\{\xi_i\}$ to estimate the integral may give two results with very different variances.

Before we can proceed to the application of Monte Carlo methods to real problems we must investigate the supply of a large number of random variables uniformly distributed on (0, 1), and the associated problem of how variables satisfying specified distributions, e.g. Gaussian, may be obtained.

7.3 Uniform Random Numbers

Monte Carlo methods depend on selecting variables x_i such that when a large number of them have been selected the histogram of the sampled distribution approximates to a specified distribution $p(x)$. In addition, the sequence in which the different values are derived should not be predetermined; there should be no correlation between the values of any member of the sequence and any other member. Variables derived in this way are *random variables*. As an example, in the study of the onset of fusion in the evolution of a star we might wish to randomly sample molecular velocities in the gas such that overall they were distributed as the Maxwell–Boltzmann distribution appropriate to the prevailing temperature. Basic to all sampling of this kind is the ability to produce values of a random variable which is uniformly distributed on (0, 1), that is $p(x)\,\mathrm{d}x = \mathrm{d}x$ for $0 \leqslant x \leqslant 1$. By definition, no mathematical formula can be given for deriving such a sequence of numbers, since all elements in the sequence would then be predetermined. Such variables can be obtained using processes occurring in nature which appear to exhibit such randomness, e.g. the distribution in time of individual cosmic rays reaching the earth, and tables of random sequences derived in this way are available. We shall say a few words about them later. Although there are cases in which truly random numbers in this sense are required, e.g. in the operation of a national lottery, in most Monte Carlo studies it is only necessary to have a sequence of numbers which has all the essential properties of a truly random sequence. Numbers which are generated following some deterministic formula and which have these properties are known as *pseudo-random* numbers. However, we shall usually refer to them simply as random numbers. Although a deterministic formula is used to produce the sequence of numbers, it is hoped that negligible correlation exists between them; if one did not know the formula one could not find any way of distinguishing between the sequence and a sequence of truly random numbers produced by some physical process such as radioactive decay. The philosophy is that if the numbers satisfy a battery of statistical tests of randomness they are acceptable irrespective of how they arose. It should be remarked, however, that workers in this field have been continuously surprised by the lack of randomness which has been revealed in sequences generated by mathematical formulae which at first encounter seemed to satisfy the criteria for randomness. Accordingly, the investigation of random number production using physical processes is by no means just a curiosity.

The best of several formulae for generating uniformly distributed random numbers is known as the *mixed congruential method*. The series of integers v_i, generated from the formula

$$v_i = (av_{i-1} + c) \bmod m \tag{7.6}$$

where a, c and m are suitably chosen non-negative numbers, form a sequence which has many of the properties of truly random numbers. The meaning of (7.6) is that v_i is m times the fractional part of $(av_{i-1}+c)/m$. An initial value v_0 (the seed) must be provided, and obviously $0 \leqslant v_i \leqslant m-1$ so that there are *at most m* distinct numbers in the sequence. In general there are not so many and, because of its deterministic nature, if any number recurs the sequence will then cycle about that number. Numbers in the range (0, 1) are readily found by taking $\xi_i = v_i/m$. Although these numbers are pseudo-random, by choosing suitable values for a, c, and m it may be possible to generate a very long sequence with 'reasonably' random properties. Such numbers have the advantage that the same sequence can always be regenerated at will and this may be helpful in repeating a calculation, e.g. where the effect of a small modification to a parameter in the problem is to be investigated. In this way, the effect of the modification can be highlighted, independent of the intrinsic stochastic variation in the system.

Two important features of such a sequence $\{\xi_i\}$ are its period, i.e. the number of variables produced before one of them recurs, and the 'degree of randomness' in it, or, conversely, the extent of correlation within the sequence. Both aspects depend on the constants a, c and m in equation (7.6). Obviously, m should be very large to get a long sequence. Typically, the largest number which can be accommodated by the word size of the computer would be used. This is $\beta^{(k-1)}$, where β is the number base of the computer and $(k-1)$ is the number of bits assigned to the representation of an integer in the machine. Extensive studies to determine optimal values for these constants have been made (for details see §3.2 of Knuth 1981) with the result that all modern computers are provided with the software to produce such a sequence, either requesting a seed, or generating one in a pseudo-random way, e.g. from the instantaneous configuration of bits representing a timing oscillator. With an optimal value for m, given by $m = 2^{(k-1)}$, a simple algorithm (Knuth 1981), which is a linear congruential generator, with well tested properties can be found by taking a in the region $(m/100)$ to m, but such that a mod $8 = 5$ and c is not a factor of m. This gives the full period m, which for $k = 32$ is about 2×10^9. In this way one can be reasonably confident that the period of the sequence will be long enough for practical Monte Carlo applications.

Any random number generator must be extensively tested for uniformity and correlations (periodicities). The uniformity of the numbers can be tested using the χ^2 test discussed in §A.3 of the Appendix. The question of the degree of correlation among the generated numbers is more difficult to assess because of the very great number of possible types of correlation.

If ξ_i is a sequence of variates drawn from a uniform random distribution, so also is the sequence $(1-\xi_i)$. Obviously the sequence $\xi_1, (1-\xi_1), \xi_2, (1-\xi_2), \ldots$, even though it would satisfy the χ^2 test for uniformity, does not conform to our idea of randomness because of strong pair correlation. Different tests can be devised to search for correlation, one of the more important ones being a test on the frequency of runs. Details are given in the books by Knuth (1981, §3.3.2) and by Bradley (1968, §12); see also Exercise 7.2.

For ideal random numbers one would expect the points defined by successive k-tuples of variates, i.e. $\pi_i^{(k)} \equiv (\xi_{k(i-1)+1}, \xi_{k(i-1)+2}, \ldots, \xi_{ki})$, to be uniformly distributed in the k-dimensional unit hypercube. Values of k-tuples like this are often used in practice to generate variables distributed as some other distribution, e.g. we will see in §7.4 how 12-tuples can be used to approximate a Gaussian distributed variable.

Sequences generated using the linear congruential method are far from being ideally randomly distributed, e.g. Marsaglia showed that if one takes $c = 0$ and $m = 2^{32}$ in equation (7.6) and considers triples of successive values, i.e. $\pi_i^{(3)} = (\xi_{3i-2}, \xi_{3i-1}, \xi_{3i})$, the totality of generated points π_i defined by these coordinates are not uniformly densely distributed in the unit cube, as would be the case for random numbers, but are confined to 2953 planes within it; this is to be contrasted with the 10^8 or more planes which would be populated by the similar total of points derived from perfect random numbers. Exercise 7.3 dramatically illustrates this very undesirable feature.

As a result of this lack of perfection in the pseudo-random numbers, small effects have been noticed in some simulations. A possible way of testing whether this is a source of error is to repeat the calculation with the complementary sequence, i.e. $(1-\xi_1), (1-\xi_2), \ldots$, and see whether statistically significant differences occur in the results. Another way would be to compare a part of the results with the results obtained using variates from a table of true random numbers.

Such tables of random numbers, derived from random physical processes, are available on magnetic tape which may be used in this way (The Rand Corporation 1955). These are processes for which we believe the probability of an occurrence in any infinitesimal time interval dt is proportional to dt. As an example of the source of such natural random numbers we may mention the randomness associated with radioactive phenomena, such as the probability of a decay occurring in a radioactive sample in a given time, or the probability of occurrence of shot noise in a thermionic valve, both of which have been used as sources for random numbers. An arrangement to obtain a sequence of random binary bits (which could then be combined to form random integers on an interval) might be to employ the Buffon needle discussed in Exercise 7.4. An alternative method, less fraught with sources of bias, would be to gate an oscillator using a signal due to cosmic ray muons. The number of oscillations is as likely to be odd as even. Assigning 0 to an odd reading and 1 to an even reading of the oscillator, a sequence of random binary digits could be built up. Such methods are slow in practice but have been used to build up libraries of random numbers which have satisfied all tests to which they have been subjected.

7.4 Random Variates from an Arbitrary Distribution

7.4.1 *Standard Methods*

It may be that we wish to produce variables distributed as the uniform distribution, e.g. the polar angle ϕ of a scattered particle in an azimuthally symmetric collision, in which case $\phi_i = 2\pi\xi_i$, but more usually we want instances of a variable x which is distributed as $p(x)$ say, which may be discrete or continuous. Henceforth ξ represents a variable drawn from the uniform distribution on $(0, 1)$.

If the variable has only two discrete values, x_1 with probability p_1 and x_2 with probability p_2 where $p_2 = (1 - p_1)$, obviously taking $x = x_1$ if $\xi < p_1$, and x_2 otherwise will reproduce the distribution $p(x_i)$. If x has three discrete values we select x_1 if $\xi < p_1$, x_2 if $\xi < (p_1 + p_2)$ and x_3 otherwise. Generalising this, we see that a variable which takes on discrete values x_i with frequencies $p(x_i)$ can be sampled by taking a value x_k where k is the smallest integer which satisfies

$$P(x_k) \equiv \sum_{i=1}^{k} p(x_i) \geqslant \xi \tag{7.7}$$

that is, x_k is the smallest value of x for which the cumulative probability equals or exceeds ξ. Important practical applications of this would be sampling from the transition probabilities W_{mn} in §7.2, or from the Poisson distribution, equation (A.15), or the binomial, equation (A.13). The Poisson distribution would, for instance, characterise the number of collisions, n, a gas molecule makes in travelling a total distance d when the mean free path is λ, and thus the mean number of collisions on such a path is $\mu = d/\lambda$;

$$p(n) = \mu^n \, \mathrm{e}^{-\mu}/n! \qquad n \geqslant 0.$$

When μ (or αN in the case of the binomial) becomes large, equation (7.7) would involve a lot of computation; the distribution may then be replaced (in practice for $\mu > 10$–20) by a Gaussian with standard deviation $\sigma = \sqrt{\mu}$. We will discuss the selection of a variate from a Gaussian distribution later.

If the quantity x is continuously distributed on (a, b) with normalised frequency $p(x)$, equation (7.7) can be generalised to obtain variables with this distribution. Solve

$$P(x) = \xi \tag{7.8}$$

where

$$P(x) = \int_a^x p(x) \, \mathrm{d}x \qquad 0 \leqslant P \leqslant 1.$$

The validity of this formula can also be seen graphically by referring to figure 7.1. By selecting uniformly from $P(x)$, we see that we are selecting regions of

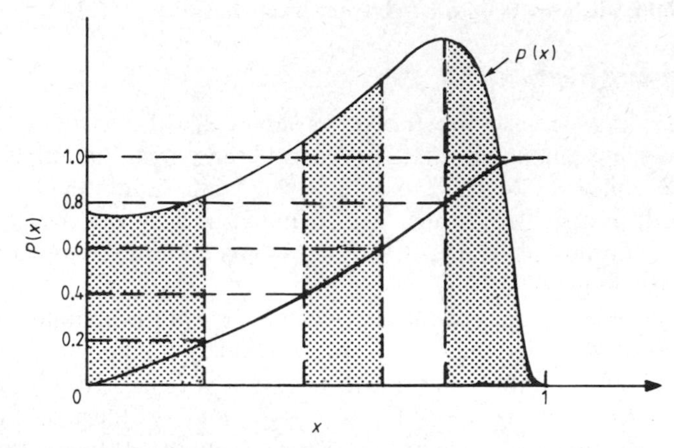

Figure 7.1 Selecting uniformly at random values of the cumulative probability $P(x)$ results in values of x being selected which are distributed as $p(x)$, a graphical justification for (7.8).

x which encompass equal areas under the distribution $p(x)$, and that is just what selecting at random from $p(x)$ requires; the probability of selecting any interval dx should be proportional to the area supported by $p(x)$ in that interval. Equation (7.8) assumes that the ξ_i values are continuously distributed; in practice they form a discrete set, e.g. via equation (7.6). In this case, only if m is large enough will the transformation (7.8) result in a sequence x_i which is not significantly different from $p(x)$. Although it may not always be a practical way for selecting from the distribution $p(x)$, because of the difficulty of solving equation (7.8) for x, it is a fundamental step in Monte Carlo work. Examples of its applications are:

(i) the exponential distribution, equation (A.9), for which

$$P(x) = 1 - \mathrm{e}^{-x/\mu} \rightarrow x_i = -\mu \ln \xi_i \tag{7.9}$$

where use has been made of the fact that $1 - \xi$ is also a uniformly distributed variable on (0, 1);

(ii) the power law distribution

$$p(x) = \left(\frac{\gamma - 1}{x_0^{\gamma - 1}}\right) x^{-\gamma} \qquad x_0 \leqslant x \qquad \gamma > 1$$

$$P(x) = 1 - \left(\frac{x_0}{x}\right)^{\gamma - 1} \rightarrow x_i = x_0 \xi_i^{-1/(\gamma - 1)}.$$

Even in cases where the equation can be solved, it may involve the evaluation of higher level functions, as in these two examples. In such a case, other ways of selecting from the distribution may be preferable for reasons of speed. The time required to generate the variate ξ on a modern computer is of the order of 2 μs so the overall selection process should not differ from this by more than an order of magnitude.

As examples of distribution functions $p(x)$ for which equation (7.8) cannot be solved explicitly, we may cite important cases like the Gaussian, equation (A.8), and the gamma distribution, equation (A.11). In such cases, methods of sampling must be found. Special methods exist for some distributions which we shall look at later. In the case of the general problem the simplest method which suggests itself is *tabular interpolation*. Divide the region of x into N (preferably a power of two) equiprobable intervals, i.e. solve once for x_i in $P(x_i) = i/N$ for $i = 0, \ldots, N$ and store the array of $(N+1)$ values of x_i. Linear interpolation can be made on this array as follows; let I and r be the integer and fractional parts of $N\xi$, i.e. $N\xi = I + r$, then choose the value

$$x = x_I + r(x_{I+1} - x_I).$$

The most important general method is *von Neumann's rejection method.* We will illustrate the method with the variable restricted to a finite interval $a \leqslant x \leqslant b$ and the distribution $p(x)$ bounded, with a maximum value p_0. We write the distribution $p(x)$ in the form

$$p(x)\ \mathrm{d}x = CQ(x)\pi(x)\ \mathrm{d}x \tag{7.10}$$

where $\pi(x)\ \mathrm{d}x$ is a distribution from which we are able to sample by one of the simple methods outlined earlier. The first part of the right-hand side of the equation is given by

$$\begin{aligned} CQ(x) &\equiv p(x)/\pi(x) && a \leqslant x \leqslant b \\ &= 0 && \text{otherwise} \end{aligned}$$

and the constant C is separated out to make the maximum value of Q equal to unity. The interval on which $\pi(x)$ is defined, and normalised, must include the interval (a, b) but need not coincide with it. The method is as follows: (i) use random number ξ_1 to select a tentative value from $\pi(x)$, call it x'; (ii) evaluate $Q(x')$ and compare it with a second uniform random variate ξ_2. If $Q(x') > \xi_2$, x' is chosen, otherwise we reject x', return to step (i) and repeat the procedure using new random numbers until the condition is satisfied. A proof of this method is given in Carter and Cashwell (1975). Its plausibility will be strengthened by considering two examples. The simple formula (7.8) cannot be used to select from the distribution

$$p(x)\ \mathrm{d}x = \frac{2x}{(2-x)^3} \exp\left(\frac{-x}{2-x}\right) \mathrm{d}x \qquad 0 \leqslant x < 2. \tag{7.11}$$

We will simply take $\pi(x)\ \mathrm{d}x = \frac{1}{2}\,\mathrm{d}x$, then

$$p(x) = 2p_0\, Q(x)\, \tfrac{1}{2}\, \mathrm{d}x$$

where $Q(x) = p(x)/p_0$ and $p_0 = 1.259$ is the maximum value of p. Referring to figure 7.2(*a*), we see that in this case $x'(\xi_1) = 2\xi_1$ and the pairs of values of (x', ξ_2) represent points uniformly distributed over the rectangle ABCD, whereas those points (x', ξ_2'), where ξ_2' satisfies $\xi_2' < Q(x')$, form the subset of the rectangle ABCD which is bounded by $Q(x)$. Hence, the probability of a point x' being selected in a region dx is just proportional to the area subtended by $Q(x)$ on dx. Obviously, the efficiency of this method, i.e. the average number of attempts to select a value for each value selected, will depend on the shape of the function $Q(x)$, being greater the closer this function comes to filling the rectangle (see Exercise 7.5). In the present case the efficiency will only be around 40%. Although use of the uniform distribution is always possible in such cases, efficiency can often be greatly improved by choosing a function $\pi(x)$ which ensures that $Q(x)$ is close to unity over much of the region. As an example we consider the distribution

$$p(\theta)\ \mathrm{d}\theta = C(m) \sin^m \theta\ \mathrm{d}\theta \qquad 0 \leqslant \theta \leqslant \pi \tag{7.12}$$

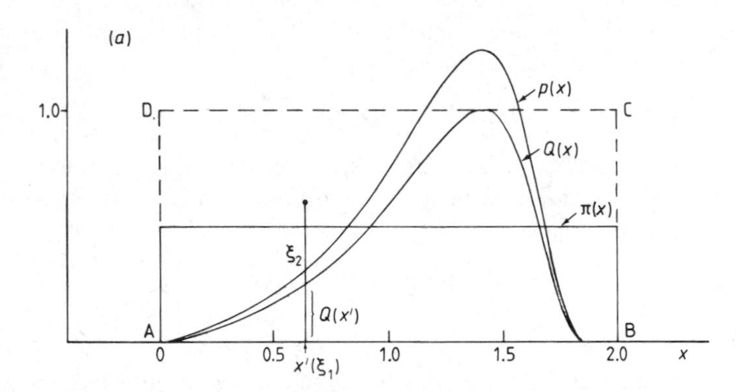

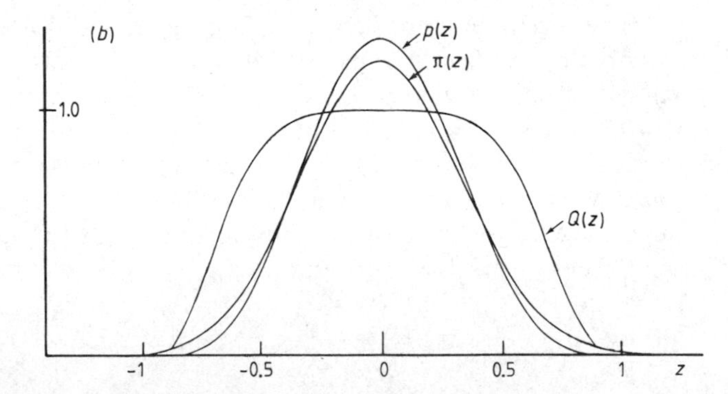

Figure 7.2 (a) Decomposition of the distribution $p(x)$ into a probability function $\pi(x)$ and a rejection function $Q(x)$ for the distribution (7.11). (*b*) Decomposition of the distribution $p(z)$ into a probability function $\pi(z)$ and a rejection function $Q(z)$ for the distribution (7.13).

which on changing to the variable $z = \cos\theta$ gives

$$p(z)\,\mathrm{d}z = C(m)(1-z^2)^{(m-1)/2}\,\mathrm{d}z \qquad -1 \leqslant z \leqslant 1$$
$$= 0 \qquad \text{otherwise.} \tag{7.13}$$

For large values of m, this is a very peaked function (see the case of $m = 10$ in figure 7.2(b)) and using the method in the previous example would be very inefficient. Instead, we write (7.13) in the form of equation (7.10) taking $\pi(z)$ as a Gaussian with variance $1/(m-1)$, that is

$$\pi(z)\,\mathrm{d}z = [(m-1)/2\pi]^{1/2}\exp\{-z^2/[2/(m-1)]\}\mathrm{d}z \qquad -\infty < z < \infty$$

in which case we have

$$Q(z) = (1-z^2)^{(m-1)/2}\exp\{z^2/[2/(m-1)]\}.$$

In this way (cf figure 7.2(b)), because our tentative sampling using a Gaussian distribution for $\pi(z)$ emphasises just those regions of $p(z)$ which are most likely, we have Q close to unity over a large region and hence a small likelihood of rejection. This is an illustration of the concept of *importance sampling* which, as we shall see later, plays a key role in the reduction of the variance in a calculation.

*7.4.2 *Markov Chain Methods*

There exists another, more abstract, method for generating members from a specified distribution based on the stationary distributions of *Markov chains*, and since it is of great importance in the application of Monte Carlo methods to problems in statistical mechanics and the quantum theory of fields, we discuss it here. Readers who are satisfied with an introduction to the subject, or whose interest will be mainly in particle transport problems, can skip the remainder of this section with no loss of continuity.

It may be that the functional form of the probability density function on $a \leqslant x \leqslant b$ is known but the normalising constant is not known, that is

$$p(x)\,\mathrm{d}x = Cf(x)\,\mathrm{d}x \qquad C^{-1} = \int_a^b f(x)\,\mathrm{d}x.$$

This could be so if the variable x itself has many components, e.g. p might be the square of the wavefunction of a system which is a function of a great number of space, spin and other coordinates. The evaluation of C requires the evaluation of a multidimensional integral, which itself might require Monte Carlo methods. The rejection technique described earlier can still be used to select from $p(x)$; however, since the search for an importance function $\pi(x)$, which by definition would resemble $f(x)$, would in general require evaluation of a similar integral to normalise it, the simplest importance function, i.e. $\pi = 1/(b-a)$, would have to be used. As we have seen, this could be very inefficient, we would have to make many attempts to select a value for x. A method based on a random walk, first described in Metropolis (1953), provides a way of producing a sequence of values of x which will follow $p(x)$.

This method has many features in common with the random walk discussed in §7.2 and depends on finding a stochastic transition operator, $W(p)$, such that starting with any initial value x_i in the region (which we will assume finite) a sequence of values is generated which are, eventually, distributed as $p(x)$. For simplicity, we take the case of a discrete distribution and as an example consider that shown in figure 7.3, where the region contains only four possible values of the variable, labelled $x_1, \ldots, x_4$. Starting from any of these points, if we execute a one-dimensional random walk among the four values, using as one-step transition probabilities the values in the stochastic matrix **W**, that is

$$\sum_j W_{ij} = 1 \qquad \text{all } i$$

where

$$\mathbf{W} = \begin{bmatrix} \frac{2}{5} & \frac{6}{25} & \frac{4}{25} & \frac{1}{5} \\ \frac{12}{25} & \frac{8}{25} & \frac{1}{5} & 0 \\ \frac{12}{25} & \frac{3}{10} & \frac{1}{10} & \frac{3}{25} \\ \frac{4}{5} & 0 & \frac{4}{25} & \frac{1}{25} \end{bmatrix}$$

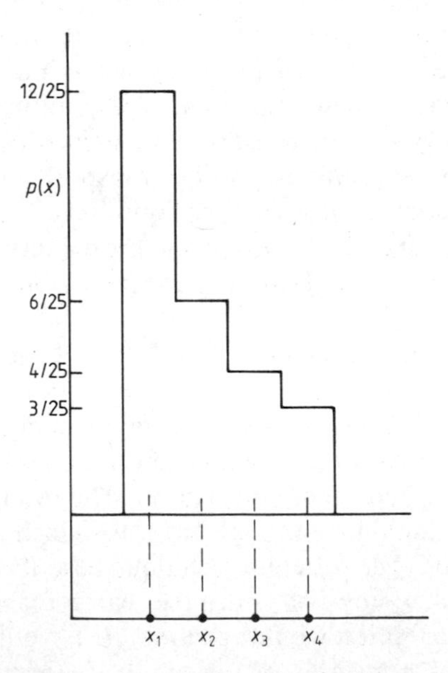

Figure 7.3 An example of a discrete distribution, with four possible values. A random walk among these values using the stochastic matrix in the text will give rise to their occurrence with the relative frequencies shown.

a sequence of values of the coordinates will evolve. By executing such a walk we mean that if, for example, we are at x_3 the next step is to remain at x_3 with probability $W_{33} = 1/10$, move to x_1 with probability $W_{31} = 12/25$, to x_2 with probability 3/10 etc. It can be shown that using this operator, no matter which point is taken as the starting point, after 'sufficient' steps all the remaining points will occur in the walk. Furthermore, their relative frequency of occurrence will stabilise to definite ratios. In fact they will be distributed as shown in the upper part of the figure, with $p(x_1) = 12/25$, $p(x_2) = 6/25$, $p(x_3) = 4/25$ and $p(x_4) = 3/25$. In the theory of Markov chains the random walk dictated by W is *ergodic*, i.e. all points are eventually accessible from all others in a finite number of steps, and the distribution $p(x)$ is the *stationary distribution* of the walk whose one-step probability matrix is $\mathbf{W}$. Not every stochastic matrix gives rise to a stationary distribution, i.e. where the memory of the initial state is lost. An important problem with Markov chains is to decide whether a given matrix does give rise to such a distribution, and how to find it. We are more interested in the inverse problem, namely given a discrete distribution $p(x_i)$ with r possible values for x_i, how do we find the one-step transition operator $W(p)$ which, with any starting value and successive application, will generate values from this distribution? The requirements are (Hammersley and Handscomb 1964)

(i) $W_{ij} \geqslant 0$ and $\Sigma_{j=1}^{r} W_{ij} = 1$ for all i, the stochastic condition;
(ii) for some finite n, $(W^n)_{ij} > 0$ for all i and j, the ergodic condition;
(iii) $\Sigma_{i=1}^{r} p(x_i) W_{ij} = p(x_j), j = 1, \ldots, r$, the convergence condition.

Since there are r^2 elements in $\mathbf{W}$, there is still flexibility in choosing the elements. A convenient further requirement—corresponding physically to microscopic reversibility—is to take

$$p(x_i) W_{ij} = p(x_j) W_{ji}.$$

Together with condition (i) this will ensure that condition (iii) is satisfied.

How rapidly the sequence generated from an arbitrary starting value converges to represent samples from the distribution $p(x)$ depends on the choice for $\mathbf{W}$. If in a single step of the walk from a point x_i, v points are directly accessible, apart from the point itself (v independent of i), then for those values of j possible elements of $\mathbf{W}$ are

$$W_{ij} = \left(\frac{1}{v+1}\right) \frac{p(x_j)}{p(x_i) + p(x_j)}$$

and

$$W_{ii} = 1 - \sum_{v \text{ values of } j} W_{ij}.$$

For example, if points on either side of a point as in figure 7.3 are accessible, $v = 2$ and W_{ii} and $W_{i,i\pm1}$ only are non-zero. Since these elements only involve ratios of the probabilities, the troublesome normalising constant C

does not enter into the method. To realise this walk in practice we separate the W_{ij} into two factors, if we are at point x_i we then choose with equal probability one of the $(v + 1)$ points, i.e. x_i itself and the v points accessible from it. If point x_i is chosen at this stage it is the next step on the walk; if not, using a uniform random deviate ξ_i, a choice among the $(v + 1)$ points is made on the basis of the probabilities

$$W_{ij} = p_j/(p_i + p_j) \qquad\qquad W_{ii} = 1 - \sum_j W_{ij}.$$

Note that the point x_i has a further chance of being chosen at this stage.

Another possible choice for **W**, usually referred to as the Metropolis algorithm (Metropolis 1953), is $W_{ij} = 0$ if state j differs from state i in the coordinates of more than one particle. We have

$$\begin{aligned} W_{ij} &= \left(\frac{1}{v+1}\right) \cdot 1 && \text{if } p(x_j) \geqslant p(x_i) \\ &= \left(\frac{1}{v+1}\right) \frac{p(x_j)}{p(x_i)} && \text{if } p(x_j) < p(x_i) \end{aligned} \tag{7.14}$$

and

$$W_{ii} = 1 - \sum_{j \neq i} W_{ij}.$$

In practice, these methods will only be used where x_i are points in a many-dimensional space, as in the phase space of a statistical mechanical system. The Metropolis algorithm is the one most commonly used in calculations in such a space and we will return to its practical realisation in Chapter 8. For a given **W**, how many steps is 'enough' will depend on the initial value and the nature of the distribution; in practice it will have to be determined on the basis of some expectation of the distribution. Further information, with examples, on Markov chains is given in §6C of Reichl (1980) and an introduction to the basic probability underlying them is given in Myhre (1970).

7.4.3 *Special Methods*

Special methods, which may be very efficient, can be used for some distributions and we will illustrate a few of these; others are given in Carter and Cashwell (1975), while optimised algorithms for many important statistical distributions are given in Chapter 9 of Fishman (1978). By far the most important is the Gaussian distribution given in equation (A.8) of the Appendix:

$$p(t)\,\mathrm{d}t = (2\pi)^{-1/2} \exp(-t^2/2)\,\mathrm{d}t.$$

The general Gaussian distribution (μ, σ) can be sampled by writing $x = \mu + \sigma t$. The importance of the Gaussian distribution arises from the fact that many physical distributions, as for example the recorded response of detectors to incident particles, can be approximated by this distribution, which must thus be included in the simulation of experimental signals. As noted, equation (7.8) cannot be solved simply in this case, but particular methods are applicable. For two independent quantities, t and u, distributed like this, that is

$$p(t)\ \mathrm{d}t = (2\pi)^{-1/2} \exp(-t^2/2)\ \mathrm{d}t$$

and

$$p(u)\ \mathrm{d}u = (2\pi)^{-1/2} \exp(-u^2/2)\ \mathrm{d}u$$

consider the joint probability distribution of finding t in $(t, t + \mathrm{d}t)$ and u in $(u, u + \mathrm{d}u)$. This is given by

$$p(t)p(u)\ \mathrm{d}t\ \mathrm{d}u = (2\pi)^{-1} \exp[-(t^2 + u^2)/2]\ \mathrm{d}t\ \mathrm{d}u.$$

Writing the transformation $t = r \cos\theta$, $u = r \sin\theta$, the right-hand side becomes

$$(2\pi)^{-1} \exp(-r^2/2) r\ \mathrm{d}r\ \mathrm{d}\theta$$

where $0 \leqslant r$, which we interpret as the joint probability density of finding the independent variables r in $(r, r + \mathrm{d}r)$ and θ in $(\theta, \theta + \mathrm{d}\theta)$, given by

$$p(r)p(\theta)\ \mathrm{d}r\ \mathrm{d}\theta.$$

Since the distribution is separable, we select r and θ separately using $P(r) = \xi_1$ and $P(\theta) = \xi_2$. In this way we have selected two independent variates from $p(t)$; $t = [2 \ln(1/\xi_1)]^{1/2} \cos 2\pi\xi_2$ or $t = [2 \ln(1/\xi_1)]^{1/2} \sin 2\pi\xi_2$. Although this method of generating a quantity distributed as a Gaussian is rigorous, it may be relatively slow to compute because of the necessity of calling up the three higher level functions log, sine and square root.

An alternative method, non-rigorous but in many cases likely to be as good an approximation to whatever phenomenon is being studied as the intitial application of the Gaussian distribution, depends on the Central Limit Theorem (§A.2 of the Appendix) and the fact that the variance of the uniform distribution is 1/12. That is, we have

$$\sigma_\xi^2 \equiv \int_0^1 \mathrm{d}\xi\ p(\xi)[\xi - (1/2)]^2 = 1/12.$$

Then for a sum of a sequence of N such variates

$$\sum = \xi_1 + \xi_2 + \cdots + \xi_N$$

an estimate of its standard deviation, which is asymptotically correct, is

$\sigma_\Sigma \simeq \sigma_\xi \cdot \sqrt{N} = \sqrt{N/12}$. Hence, a variable defined as

$$t = \sqrt{12N}\left(\frac{1}{N}\sum_{i=1}^{N} \xi_i - 0.5\right)$$

has a mean of zero and a variance of 1. A convenient, and more than sufficiently large, value of N is 12, in which case the variable

$$t = \sum_{i=1}^{12} \xi_i - 6 \tag{7.15}$$

will be distributed with mean zero and variance 1, and will be approximately Gaussian. Although it requires 12 random numbers to generate, it will in general be faster than the method described earlier and is a commonly used method. However, in view of the non-uniformity of triplets noted above, some reservation seems necessary.

Often the sine or cosine of a uniformly distributed polar angle is required; the rejection method discussed above leads to the following algorithm for their selection:

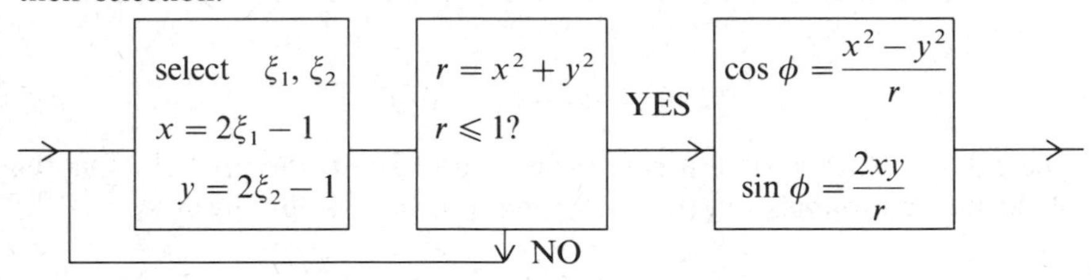

Of the initial points x, y uniformly distributed over the unit square, only those which fall inside the unit circle are accepted. Because the annulus of area is proportional to $\Delta\phi$, these points select ϕ uniformly distributed on $(0, 2\pi)$; the final expressions follow from trigonometry. The method has an efficiency of $\pi/4$ and the merit of not calling up any high level functions.

Some other important special methods are described in the exercises at the end of the chapter.

7.5 Particle Transport

As an example of the application of the ideas discussed so far in this chapter we outline some of the methods used in particle transport calculations. These may arise from many different situations of which the following are typical examples from physics:

(i) the investigation of the cascade produced by a high energy electron, photon, or hadron in an ionisation calorimeter used at accelerators to estimate the energies of such particles;

(ii) studies of similar cascades produced by primary cosmic rays incident on the earth's atmosphere to estimate the total amount of Cerenkov radiation which is emitted;

(iii) a calculation of the penetration of a beam of energetic γ-rays through a block of material to determine its effectiveness as a shield;

(iv) a study of the expected evolution of a configuration of neutrons in a nuclear reactor assembly where the competition between absorption and fission will determine the behaviour.

Although the quantity sought, i.e. the 'score' in each case, will be quite different, e.g. the energy deposited by ionisation at different depths in the calorimeter, the distribution in intensity of the resulting Cerenkov light at ground level in the case (ii), the average number of particles which escape per incident particle in the shielding problem, and the energy density released per unit time in the reactor, the solutions have much in common. This can be illustrated by reference to figure 7.4 where we show, schematically, a possible sequence of events when an energetic cosmic ray nucleon strikes the atmosphere. Different particles will be created, each of which can undergo certain physical processes, e.g. photons may undergo Compton scattering on an electron in the material, process (a), or, if energetic enough, may give rise to an e^+e^- pair, process (b). Electrons may produce γ-rays by bremsstrahlung, process (c). High energy hadrons in a collision will produce a multiplicity of secondary particles, e.g. π^0 mesons, which decay rapidly to $\gamma\gamma$, process (d), charged pions which may decay to a muon, process (e), or collide with nucleons to produce further secondaries, process (f). In the case of fission, the secondaries will be neutrons whose fate may be absorption (which may also happen to low energy π^-, process (g)). Depending on the nature and energies of the particles involved, many of these processes may be of negligible importance; several others, such as positron annihilation or photon absorption by the photoelectric effect, may occur but in any case the general

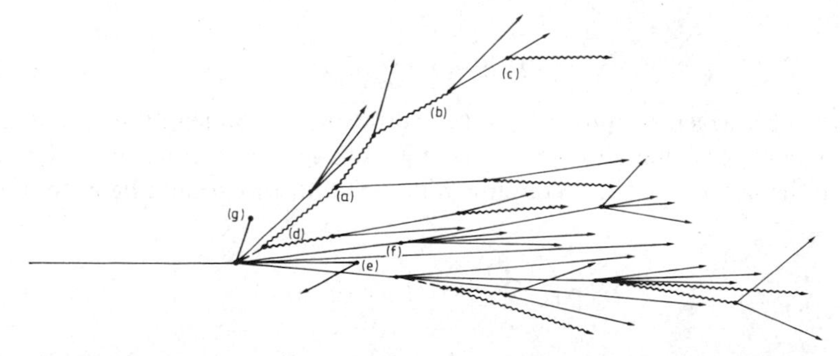

Figure 7.4 An illustration of some of the phenomena occurring in the interaction of a cosmic ray nucleon in the Earth's atmosphere. Among possible processes are Compton scattering, (a); pair production, (b); bremsstrahlung, (c); π^0 decay to $\gamma\gamma$, (d); $\pi \to \mu\nu$, (e); charged pion interaction, (f); π^- absorption, (g).

'tree' structure of the transport will be obtained. The positions of the nodes on the tree are determined by some functions (cross sections) of the parameters of the incoming particle and the composition of the medium in each case, and will have to be sampled from the appropriate distributions. Many other such samplings will have to be made at each node to determine the multiplicity and nature of the outgoing particles, their direction of motion, energy etc. When more than one particle comes out from a node, i.e. an inelastic interaction, all but one must be stored, since only one can be followed at a time. To keep the storage requirements to a minimum it is advisable to process first the particle which is likely to produce the least complicated branch; in general it will be the secondary with lowest energy. Secondaries from an interaction should thus be stored such that they will be accessed in order of increasing energy. It was ideas like these which inspired the modern development of Monte Carlo methods: for an interesting application before the age of the electronic computer—using a mechanical drum for event selection—you should consult Wilson (1952).

The probability of a particle interacting in a path $\mathrm{d}l$ may be written as $\mu\,\mathrm{d}l$, where the coefficient μ may depend both on the particle (including its energy) and the nature of the medium and may have contributions from different physical processes.

We now outline some of these considerations where the incident particle is a photon of energy E_γ, important, for example, in shielding calculations or in radiobiological simulations to estimate radiation dosage. In this case, relevant processes are the photoelectric effect, Compton scattering and pair production, i.e. $\mu = \mu_\mathrm{p} + \mu_\mathrm{C} + \mu_\mathrm{pp}$. In general, $\mu_i = n\sigma_i(E)$ where σ_i is the cross section for the process and n is the density of targets. For example, for Compton scattering $n = Z\rho N_A/\mathscr{M}$ where ρ is the density, $\mathscr{M}$ the molecular weight of the medium, Z the number of target electrons per atom and N_A is Avogadro's number. The probability that a photon has its next interaction in $\mathrm{d}l$ at l is

$$p(l)\,\mathrm{d}l = \mathrm{e}^{-l/\lambda}\,\mathrm{d}l/\lambda \qquad (7.16)$$

where $\lambda(E)$, which is equivalent to $1/\mu$, is the interaction length. For particles whose energy is changing gradually, e.g. through ionisation, or in the case where the density of the medium is not constant, this should be generalised to

$$\exp\left(-\int_0^l \mathrm{d}l'/\lambda(l')\right)\left(\mathrm{d}l/\lambda(l)\right).$$

When an interaction node is located by selecting from (7.16), the particular type of interaction must be selected with relative probability $\mu_\mathrm{p}(E)$: $\mu_\mathrm{C}(E)$: $\mu_\mathrm{pp}(E)$. By way of example, we will assume Compton scattering has been selected; consideration of this process is important not only in shielding problems but also, for example, in the design of a satellite-borne detector of cosmic x-rays, which depends on the Compton effect. We must then select directions and energies for the outgoing electron and photon (see figure 7.5). In this case, because there are only two outgoing particles, the scattering angle θ of the photon and its new energy, E'_γ, are uniquely related and given by

$$\frac{E'_\gamma}{m_e c^2} = \frac{\varepsilon}{1 + [\varepsilon(1-q)]} \tag{7.17}$$

where $\varepsilon \equiv E_\gamma / m_e c^2$ and $q = \cos\theta$. The distribution of the photon's energy, in terms of z ($z \equiv E'_\gamma / E_\gamma$), is given by the Klein–Nishina formula, see §22 of Heitler (1954),

$$p(z)\,\mathrm{d}z = C(\varepsilon)\,[z + (1/z) + q^2(\varepsilon, z) - 1]\,\mathrm{d}z \qquad 1/(2\varepsilon+1) \leqslant z \leqslant 1 \tag{7.18}$$

where, from (7.17), $q(\varepsilon, z) = 1 + (1/\varepsilon) - (1/\varepsilon z)$ and $C(\varepsilon)$ is a normalising constant. Rejection techniques of the type discussed in §7.4 (and Exercises 7.8 and 7.16 at the end of the chapter) can be used to select a value for $E' = zE$, and thus, via (7.17), for $\cos\theta$. The azimuthal angle which is uniformly distributed, or its sine and cosine, may be selected as described in §7.4 and hence the energy and direction cosines of the outgoing photon are determined. The corresponding quantities for the electron are determined by the kinematics,

$$E_e = (1-z)E_\gamma \qquad \tan\phi = (1+\varepsilon)^{-1}\cot(\theta/2).$$

One of these particles will then be stored and the other followed. This must be continued until all particles have come to rest, or been absorbed, or escaped from the shield. If a sample of N photons incident on the block is treated in this way and a total of n photons or electrons escape, the estimate of the transmission coefficient T is simply $T = n/N$. Although formally a straightforward calculation, as we will now see, considerable refinement will be necessary for its application in practice.

The number n is a variate from a binomial distribution (equation (A.13) of the Appendix) where the probability of success is $\alpha = T$. Hence, there is a variance associated with it, $\sigma_n^2 = NT(1-T)$. The corresponding standard

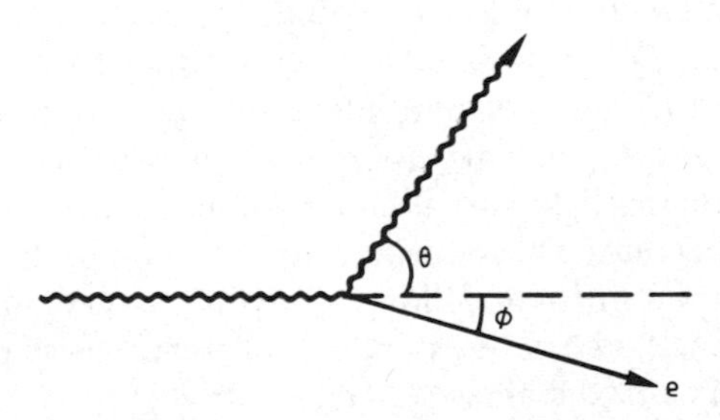

Figure 7.5 The geometry of Compton scattering by an electron.

deviation on T is $\sigma_T = \sigma_n/N = [T(1-T)/N]^{1/2}$. In practice, for a shield where $T \ll 1$ (n has a Poisson distribution), it is found that $\sigma_T = \sqrt{T/N}$, which, using our estimate of T, gives $\sigma_T = \sqrt{n}/N$. If we require this standard deviation on the result not to exceed 10% of T, say, then we must have $\sqrt{n}/N \lesssim 0.1\ n/N$, i.e. $n \gtrsim 100$. If the actual transmission coefficient T is of the order of 10^{-6}, a desirable value for good shielding, the number of photons N which must be tracked in the shield is 10^8 (where $N = n/T$). Each of these individual events, in turn, involves tracking a great number of secondaries, the great majority of which, as distinct from the ionisation calorimeter calculation, make no contribution to the score n. Simulation of particle histories in this way, where the particle configurations are a direct analogue of a physical situation, is called an analogue Monte Carlo calculation. Various approaches have been suggested to increase the quantity of information in such calculations, and such methods, known as non-analogue Monte Carlo, are examples of general methods of variance reduction which we will now consider.

*7.6 Variance Reduction Techniques

In the analogue Monte Carlo calculation of penetration through a shield with a small transmission probability, as discussed above, the result of tracking a primary particle, in most cases, will be that no particle exits at the other end. We seek some way of emphasising those cases where some particles do exit and thus contribute to the statistics on the score; at the same time we must not introduce any bias into the resulting answer. This may be achieved by distorting the distributions selected from and attaching weights to individual particles in the cascade which correct for this distortion and ensure conservation of all conserved quantities. A brief description of some specific methods by which this is done will illustrate the concept.

Survival biasing is a modification in which absorption of a particle is not permitted. In every interaction only those processes which do not lead to absorption of the incoming particle are sampled, but all particles emerging from the interaction selected would be assigned a weight w where $w = 1 - \mu_{abs}(E)/\mu(E)$, which is less than or equal to one, and μ_{abs} and μ represent the absorption and total interaction probabilities respectively. In the case of photons, μ_{abs} might represent the photoelectric effect. Similarly, when the incident particle carries a weight w entering an interaction the product of weights must be assigned to the outgoing particles. In this way all produced particles contribute, although possibly a very small amount, to the final score. The method can be extended to de-emphasise particles travelling away from the axis of the absorber etc.

A variation of this is von Neumann's *splitting* technique. In the penetration calculation, at one or more depths in the shield each particle crossing that level would be replaced by ν particles all having identical properties, energy, direction etc, to the incident particle but each carrying its weight divided by ν, i.e. a weight $1/\nu$ if the incident particle has a weight of unity. This will obviously be helpful in cases where a particle survives with a rather high energy at a great depth in the shield but may still only have a relatively low probability of giving rise to a transmitted particle. This splitting technique can be incorporated in a systematic way into a calculation but obviously cannot be arbitrarily frequent. If after a distance x we split ν times, the fraction of events starting from there on average is $\nu e^{-\mu_{abs}x}$, which after another distance x becomes $\nu e^{-2\mu_{abs}x}$. Splitting again gives $\nu^2 e^{-2\mu_{abs}x}$ tracks. If this is continued for j equal intervals, we will have, on average, $(\nu e^{-\mu_{abs}x})^j$ events to process. This average should not become significantly greater than unity in order not to bias in favour of the particular beginning to this chain, i.e. we require $\nu e^{-\mu_{abs}x} \lessapprox 1$. If the splitting is carried out after every absorption mean free path, i.e. $x = \mu_{abs}^{-1}$, a suitable value for ν is 2. Care must be taken when such weighted particles are scattered backwards across a splitting level to remove the weighting. The method is somewhat empirical, but with these kinds of parameters it agrees with analytic results where comparisons can be made.

Both survival biasing and von Neumann splitting methods also find important application in the simulation of non-intersecting polymer chains, see §8.6.

The opposite of splitting is *Russian roulette* which is designed to reduce the work spent on particles which a priori are unlikely to contribute much to the final score, e.g. numerous low energy particles in the early stages of a cascade. The philosophy is that at any stage of the calculation the number of particles of any energy being processed is proportional to the expected contribution of such particles to the final result. At some level, particles are discarded from future tracking with a probability α, but the fraction $(1-\alpha)$ which are continued are each assigned a weight $(1-\alpha)^{-1}$. In the shielding problem α would be a function of energy so that all high energy particles would be followed but only a fraction of the, very numerous, low energy ones would. These different methods can be combined, e.g. Russian roulette, as described, could be played to avoid wasting time on following particles which arose from splittings and whose weight has become so small that they can only make an insignificant contribution to the score. If the weight of a particle, w, falls below some specified w_{min}, compare $w/0.5$ with a variate ξ from the uniform distribution. If $\xi < w/0.5$, reassign the particle a weight 0.5, otherwise terminate it. All of these methods ensure that the score, i.e. the number of escaping particles, form a continuum, and for the same number of trials the variance on the score will be less than for integer outputs. When combined in an optimum way they can lead to a great reduction in the amount of work necessary to establish the result. If the magnitude of intrinsic fluctuations in the physical problem is part of the investigation, such weighting methods may not be suitable.

The methods described above are illustrative of a particular class of method called *importance sampling*; other methods such as *correlated sampling* and *anticorrelated sampling* exist, but since the details of all such methods are rather problem specific they will be best described in the context of the general problem of variance reduction, discussed in detail in Hammersley and Handscomb (1964). As has been pointed out earlier (§7.2) any Monte Carlo calculation can be viewed as estimating the value of a multidimensional integral. However, to simplify the notation we will present the discussion in terms of the simple integral in one variable,

$$I = \int_0^1 f(x)\, \mathrm{d}x. \tag{7.19}$$

An estimate of this integral, $\eta(\xi_1 \ldots)$, can be obtained using a sequence of one or more random variables, and repeated estimates would give a distribution of η values, known as the sampling distribution, $p(\eta)$. The expected value of the mean of this distribution in η, in the absence of bias, would be the value I, and, accordingly, a best estimate of the value I would be the mean of the distribution of η values obtained, that is

$$\int \eta p(\eta)\, \mathrm{d}\eta. \tag{7.20}$$

The variance of the distribution

$$\sigma_\eta^2 = \int (\eta - I)^2\, p(\eta)\, \mathrm{d}\eta \tag{7.21}$$

would be a measure of the spread in possible values obtained; the estimation of σ_η^2 is discussed in the Appendix. In particular, the standard error on the estimate (7.20) would be $\sigma_\eta/\sqrt{N}$, where N is the number of samples made, so that for a given number of samples the error will be less the smaller the variance on an individual estimate. As we have seen in the particle transport problem above, random variables may be used in different ways which, although they give unbiased estimates of I, may give very different values for the variance, and thus for the uncertainty on the estimate.

It has been pointed out before that a possible estimate of the integral (7.19) is the crude Monte Carlo estimate $\eta^{(1)}$ obtained by selecting an x value at random from the uniform distribution, $x = \xi$, and taking $\eta^{(1)} = f(x)$. Then, since $p(\eta)\, \mathrm{d}\eta = p(x)\, \mathrm{d}x$, the variance on $\eta^{(1)}$ is, using (7.21),

$$\sigma_{\eta^{(1)}}^2 = \int_0^1 (f(x) - I)^2\, \mathrm{d}x. \tag{7.22}$$

An alternative approach is to write (7.19) in the form

$$I = \int_0^1 \frac{f(x)}{\pi(x)}\, \pi(x)\, \mathrm{d}x \tag{7.23}$$

where $\pi(x)$ is a normalised probability distribution on (0, 1). Equation (7.23) simply says that the quantity $\eta^{(2)}(x) = f(x)/\pi(x)$, when the variable x is selected from the distribution $\pi(x)$, has the required integral I as its expectation value. The variance on this estimator is

$$\sigma^2_{\eta^{(2)}} = \int \left(\frac{f(x)}{\pi(x)} - I \right)^2 \pi(x)\, \mathrm{d}x \tag{7.24}$$

and it may be quite different from (7.22). In particular, this variance could be zero if we choose $\pi(x) = f(x)/I$, but since we don't know I this is not possible. However, this observation suggests that by a suitable choice of $\pi(x)$ we could make the variance small; this would be the case if $f(x)/\pi(x)$ was close to a constant everywhere. In other words, the function $\pi(x)$, from which we select the values of x for estimating the integral, should mimic the function $f(x)$ being integrated. In this case, values will be selected predominantly in the region where $f(x)$ is making its greatest contribution to the integral. This is the concept of importance sampling discussed earlier in §7.4, emphasising the selection of variables in regions which contribute most to the final answer, but correcting this distribution in selection by attaching weights. In this way, N random numbers from the distribution $\pi(x)$ would be used to estimate (7.23) as

$$\frac{1}{N} \sum_{i=1}^{N} w(x_i) f(x_i)$$

with $w(x) = 1/\pi(x)$. The case of splitting in particle transport corresponds to π being taken as a combination of one or more piecewise, constant step functions in one or more variables, e.g. $\pi(E, x, \theta)$; more general functions, notably the exponential function, are also used in practical calculations.

It may be that the problem to be solved is closely related to some other problem, e.g. due to a perturbation, and it is the consequences of this perturbation which are of interest. Associated with the solution of either problem there would be a variance due to the stochastic nature of the method which might mask the small difference in the solutions. To reduce this, the same set of random numbers might be used, as far as possible, to solve the two problems, in which case the significance of their difference will be greater because of the approximate cancellation of random contributions. Alternatively, it may be that one of the problems can be solved exactly; using our one-dimensional integral as an example, we could rewrite (7.19) as

$$I = \int_0^1 F(x)\, \mathrm{d}x + \int_0^1 g(x)\, \mathrm{d}x$$

where $g(x) = f(x) - F(x)$ and $F(x)$ is a function for which we can evaluate the integral exactly, say with value I_0. Then $I_0 + \eta$ is an estimate of I, where η is an estimate of the second integral, and the only variance on the answer arises from the variance on η. If the term $\int F(x)\,dx$ contains the main part of the answer, the variance on the second integral (even if calculated using crude Monte Carlo) may, relatively, be quite small, although of course variance reduction might be used in its evaluation. This method, where we find an estimate for a problem which is strongly correlated with the one in hand, is known as the method of *correlated sampling*. In the case of a high energy photon incident on a semi-infinite medium, the average number of particles crossing any level can be found analytically. This solution only differs from that of the shielding problem by virtue of the fact that it includes contributions from particles which, at some stage, are reflected backwards from beyond the end of the shield, and hence is a valuable control variate in the shielding calculation.

Just the opposite approach, where we seek two estimates of the solution which are negatively correlated, is also a powerful scheme for reducing the variance and is known as the method of anticorrelated or *antithetic variables*. It relies on the fact that the variance of a quantity z, which is itself the sum of other variables x_i, i.e. $z = \Sigma\, x_i$ has the form, cf §A.1,

$$\sigma_z^2 = \sum_i \sigma_{x_i}^2 + 2 \sum_i \sum_{j<i} \operatorname{cov}(x_i, x_j)$$

where 'cov' is the covariance. If we have two distinct methods which give estimates $\eta^{(1)}(\xi)$ and $\eta^{(2)}(\xi)$ for I in (7.19), each characterised by its variance, then a third estimate, $\eta^{(3)}(\xi) = \frac{1}{2}(\eta^{(1)} + \eta^{(2)})$, has a variance

$$\sigma_{\eta^{(3)}}^2 = \tfrac{1}{4}\sigma_{\eta^{(1)}}^2 + \tfrac{1}{4}\sigma_{\eta^{(2)}}^2 + \tfrac{1}{2}\operatorname{cov}(\eta^{(1)}, \eta^{(2)}).$$

If the covariance is negative, i.e. the estimates $\eta^{(1)}$ and $\eta^{(2)}$ are negatively correlated, this may give rise, for the same amount of work, to a value which is smaller than the variance on either estimator separately. As an illustration, we note that if the function f in (7.19) is monotonically varying over the region of integration, then two crude Monte Carlo estimates $\eta^{(1)} = f(\xi)$ and $\eta^{(2)} = f(1-\xi)$ will have this property, since if $f(\xi)$ is less than I, the mean value of f over the interval, then $f(1-\xi)$ will be greater and vice versa. In the trivial case of $f(x) = ax + b$, this situation would yield the exact answer for every estimate, formally because $\sigma^2(\xi) = \sigma^2(1-\xi) = -\operatorname{cov}(\xi, 1-\xi) = 1/12$ and hence the variance vanishes. A generalisation of this (Hammersley and Handscomb 1964) is to take

$$\eta = \alpha f(\alpha\xi) + (1-\alpha) f[1-(1-\alpha)\xi]$$

where α is a solution of $f(\alpha) - \alpha(f(0) - f(1)) = f(1)$. This method has been used in neutron transport calculations where pairs of neutrons are chosen leaving a node such that they are travelling in exactly opposite directions.

Much study has gone into the properties of these and other techniques for variance reduction; however, in the words of Kahn 'the greatest gains in variance reduction are often made by exploiting specific details of the problem rather than by routine application of general principles'. In any case, this summary should make clear that any large scale Monte Carlo programming must be preceded by careful planning and formulation.

The evaluation of an integral has been used above to illustrate the idea of variance reduction; this, however, is of more than academic interest because an important application of Monte Carlo methods is the explicit evaluation of phase space integrals in high energy physics. These integrals have the form (Byckling and Kajantie 1973)

$$I = h^{-3N} \int \prod_{i=1}^{N} \mathrm{d}^3 r_i \left(\frac{\mathrm{d}^3 p_i}{2E_i}\right) \delta^4\left(p - \sum_{i=1}^{N} p_i\right)$$

where N is the number of particles in the system. It is interesting to see why conventional methods of integration, e.g. Simpson's rule, are not competitive with the Monte Carlo method for large values of N. One way of evaluating the integral $\int_0^1 f(x)\,\mathrm{d}x$ is the 'hit-or-miss' method, essentially the acceptance–rejection method used to select numbers from a distribution described in §7.4. N pairs of random numbers $(\xi_1^{(i)}, \xi_2^{(i)})$ are selected, and the number of times, ν, that the condition $f(\xi_1^{(i)}) > \xi_2^{(i)} f_{\max}$ is satisfied is noted, this is just the number of times the point defined by $(\xi_1^{(i)}, \xi_2^{(i)} f_{\max})$ falls under the curve. An estimate of I is then ν/N and, since ν is binomially distributed, the standard deviation on this estimate is given by equation (A.14) as $N^{-1}[\nu(1-\nu/N)]^{1/2}$. For a specified accuracy this method is even less efficient than any of the methods we described earlier. However, we notice that for the same accuracy the work required, i.e. the size of N, does not depend on the dimensions of the space over which the integral is carried out. We can contrast this with conventional methods of numerical integration where if n points are used to estimate a one-dimensional integral to a specified accuracy, n^k points must be used to estimate a k-dimensional integral to the same accuracy. We can thus expect stochastic methods of integration to become viable when $n^k > N$. In practice, recourse is often made to Monte Carlo methods if $k \gtrsim 4$.

Exercises

7.1 By replacing the time derivative in (3.37) by $(U(x,t) - U(x,t-\tau))/\tau$, and using the notation in figure 3.7 and §7.2, show that suitable transition probabilities for a solution to this diffusion equation using a random walk on a rectangular grid, $h \times \tau$, are: $W_{01} = W_{03} = (2 + \lambda h/\kappa)^{-1}$, $W_{02} = 0$, $W_{04} = (1 + 2\kappa/\lambda h)^{-1}$.

7.2 The numbers $x_{i+1}, \ldots, x_{i+s}$ in the sequence $x_{i-1} > x_i < x_{i+1} \ldots < x_{i+s} > x_{i+s+1}$ are said to form a run of length s where $s \geqslant 1$. As a test of a random number generator use it to produce a large sample, N, of variates and find the frequency of runs of different lengths.

Compare the frequency distribution in s with that expected from a sequence of N random numbers, that is

$$p_N(s) = 2N(s^2 + 3s + 1)/(s + 3)!$$

7.3 As an illustration of the clustering of points which multiplets of otherwise apparently random numbers can give rise to, use (7.6), with $a = 57$, $c = 1$, $m = 256$ and $x_0 = 10$, to generate all possible numbers. Use successive pairs of these numbers, (v_{2i}, v_{2i+1}), $i = 0, 1, \ldots$, to define points in the plane and make a plot of them. For other examples which give unsatisfactory results, and a general discussion of such correlations, see the paper by Ripley (1983).

7.4 One way of testing a sequence of numbers for hidden correlations is to use the sequence to solve a problem with a known solution. Use your random generator to investigate a two-dimensional random walk of N steps on a square lattice of unit step length where the transition probabilities at every point in all four directions are equal. For each of several values of N find the root mean square distance from one end to the other, averaged over many walks. Compare your results with the theoretical expectation that, at large N, this quantity varies as $N^{1/2}$.

Repeat the above exercise for a similar walk but without self-reversal, i.e. at each step the transition probability is 1/3 for all directions except the one from which the walk has most recently come, and compare the result with the case above. Random walks like this will play an important role in our discussion of polymer chains in Chapter 8. A thorough treatment of them is given in Chandrasekhar (1943).

7.5 An interesting variant of the one-dimensional random walk in Exercise 7.4 is to imagine each step as introducing a physical rod of unit resistance, i.e. a segment which has been traversed n times has a resistance $1/n$. Find the average total resistance, $R(N)$, from the beginning to the end-point of a walk of N steps, and the distribution in this quantity. In analytical terms, this is a much more difficult problem than that of the length of the chain. Investigate it for values of $N < 500$. Another quantity which can be investigated is the resistance between the extreme ends of a walk of N steps.

Such calculations are relevant in the study of fluid permeability in rock; details, including higher dimensional cases, are given in Banavar *et al* (1983).

7.6 One of the earliest uses of the Monte Carlo method was the experimental determination of the quantity π by the Count de Buffon in the eighteenth century. If a needle of length L is thrown in an unbiased way on to a grid of lines forming squares of side d, the mean number of lines intersected is $\langle n \rangle = 4L/\pi d$. Devise a real experiment, or a computer simulation, to use this expression to estimate π. Apply the method of antithetic variables, described in §7.6, to the number of intersections on the x and y grids separately (each of which has a mean $2L/\pi d$) to find an alternative estimate of π and compare the standard error on the estimate in the two cases. For more information see the paper by Kahan (1961).

7.7 Referring to (7.10), show that the efficiency of the rejection method, i.e. the ratio of successful samples to attempts, is C.

7.8 Show that the rejection method based on (7.10) can be generalised as follows: write

$$p(x)\,\mathrm{d}x = \sum_{i=1}^{n} C_i Q_i(x)\pi_i(x)\,\mathrm{d}x$$

where the π_i are probability density functions, $C_i > 0$ and $0 < Q_i \leqslant 1$ in the region. Choose an integer k with probability $C_k(\Sigma_{i=1}^{n} C_i)^{-1}$ and then select from $Q_k\pi_k$, as described in §7.4.

7.9 The probability that the time interval between two random events lies in $(t, t+\mathrm{d}t)$, $p(t)\,\mathrm{d}t = \mu\mathrm{e}^{-\mu t}\,\mathrm{d}t$, leads to a Poisson distribution for the number of events occurring in a time T (see equation (A.15) of the Appendix) with $\bar{n} = \mu T$. Use this to select variables from the Poisson distribution as follows. Set $\mu = \bar{n}$ and select a time t_1 from $p(t) = \mu\mathrm{e}^{-\mu t}$, if $t_1 > 1$ set $n = 0$; otherwise select a second time t_2 and check if $t_1 + t_2 > 1$, if *yes* $n = 1$, if *no* continue, i.e. find the smallest k such that $\Sigma_{i=1}^{k} t_i > 1$ and set $n = (k-1)$.

Show that this may be formulated as follows: to select an n choose the smallest integer k such that $\Pi_{i=1}^{k+1} \xi_i < \mathrm{e}^{-\mu}$ where the ξ_i are from the uniform distribution on (0, 1). Write a program to compare the efficiency of this method relative to the simple method using (7.7).

7.10 In the case of the standardised normal distribution (A.8) show how values of $|t|$ can be selected using the rejection technique, with $\pi(t) = \mathrm{e}^{-t}$, and prove that the efficiency of the method is $\sqrt{2e/\pi} \simeq 0.76$.

7.11 Explain why the two normal variates selected by the method described in §7.4 are independent.

7.12 The rejection technique can sometimes be used on a variable with infinite range by transforming it to a finite range, selecting from the transformed variable and inverting the transformation to get the sample variate. Consider the Maxwell–Boltzmann distribution for molecular velocities,

$$p(v)\,\mathrm{d}v = Cv^2\,\mathrm{e}^{-\alpha v^2}\,\mathrm{d}v$$

and rewrite it in terms of a variable u where $\alpha v^2 = [u/(1-u)]^{\beta}$, with β a constant. Show that an optimum value for β is 2/3 and find the efficiency for sampling from it. For a more efficient method in this case see Chapter 9 of Fishman (1978).

7.13 Show that the Lorentz, or Breit–Wigner, distribution

$$p(x) = \frac{\Gamma(\pi)^{-1}}{(x - x_0)^2 + \Gamma^2}$$

can be selected from $x_i = x_0 - \Gamma \operatorname{cotan}(\pi\xi_i)$.

7.14 The normalised black-body radiation frequency spectrum is

$$p(x)\,\mathrm{d}x = \frac{15}{\pi^4}\left(\frac{x^4}{\mathrm{e}^x - 1}\right)\mathrm{d}x$$

where $x = h\nu/kT$. Show that the following procedure selects from it (Carter and Cashwell 1975) and find its efficiency: let L be the minimum value of the integer l for which

$$\sum_{j=1}^{l} (1/j^4) \geqslant \xi_1\pi^4/90$$

then set $x = -(1/L)\ln(\xi_2\xi_3\xi_4\xi_5)$, where all the ξ_i are from the uniform distribution on (0, 1).

7.15 Consider the random walk in §7.4 where, in addition to itself, each point is only directly accessible from the points on either side of it and show that the Metropolis algorithm leads to a one-step transition matrix, given by

$$\begin{bmatrix} \frac{3}{4} & \frac{1}{6} & 0 & \frac{1}{12} \\ \frac{1}{3} & \frac{4}{9} & \frac{2}{9} & 0 \\ 0 & \frac{1}{3} & \frac{5}{12} & \frac{1}{4} \\ \frac{1}{3} & 0 & \frac{1}{3} & \frac{1}{3} \end{bmatrix}$$

Starting with x_4, carry out a random walk using this matrix, and the one in the text, and compare the convergence to the distribution given in figure 7.3 for the two cases.

7.16 Show that the following is a possible decomposition of the Klein–Nishina distribution, equation (7.18), suitable for use with the method described in Exercise 7.8:

$$C_1 = \frac{2\varepsilon + 1}{2\varepsilon + 9} f(\varepsilon) \qquad Q_1(z) = 4(z - z^2) \qquad \pi_1(z)\,\mathrm{d}z = \mathrm{d}z/2\varepsilon z^2$$

$$C_2 = \frac{8}{2\varepsilon + 1} f(\varepsilon) \qquad Q_2(z) = \frac{z + q^2}{2} \qquad \pi_2(z)\,\mathrm{d}z = \frac{2\varepsilon + 1}{2\varepsilon}\,\mathrm{d}z$$

where $f(\varepsilon)$ is a function whose value is not required in the selection process.

Using the explicit form for $f(\varepsilon)$ it is possible to modify the method so as to avoid the initial stage of process selection, where the total probability integrated over z must be known, e.g. $\mu_C(\varepsilon)$. Investigate this. See Butcher and Messel (1960).

7.17 Referring to the photon transport problem described in §7.5, it can be shown that the average number of photons at a depth x with energy in $(E, \mathrm{d}E)$ and travelling in an element of solid angle $\mathrm{d}\Omega$ at an angle θ to the incident direction, $\gamma(x, E, \cos\theta)$, satisfies the transport equation

$$\cos\theta \frac{\partial\gamma}{\partial x} + \gamma/\lambda(E) = \iint K(E, \theta; E', \theta')\gamma \, \mathrm{d}E' \, \mathrm{d}\cos\theta'$$

where K describes the propagation in phase space from (E', θ') to (E,θ). Show that the *exponential transformation* $\gamma^* = \mathrm{e}^{cx}\gamma$, where c is a constant, gives rise to a similar transport equation for γ^*, with the interaction length replaced by $\lambda^* = \lambda(E)/(1 - c\lambda(E)\cos\theta)$. This means that in the simulation interaction points may be selected from the distribution

$$p(l)\,\mathrm{d}l = \mathrm{e}^{-l/\lambda^*} \frac{\mathrm{d}l}{\lambda^*}$$

(provided c is chosen greater than the minimum value of $\lambda(E)$ so that λ^* remains positive). This has the effect of greatly increasing the survival probability of the photons, and at the same time de-emphasising those not travelling in the forward, $\theta = 0^0$, direction. As a result, many more will exit from the shield, i.e. $\gamma^*(L) \gg \gamma(L)$, where L is the thickness of the shield. Show that, in this case, it is not necessary to make any modification to the weights of the particles; if T^* is the transmission coefficient found in this way the true transmission coefficient is simply $T = \mathrm{e}^{-cL}\,T^*$. For more details see, for example, §25 of Fano *et al* (1959).

References

Banavar J R, Harris A B and Koplik J 1983 *Phys. Rev. Lett.* **51** 1115–8

Braddick H J J 1966 *The Physics of Experimental Method* 2nd edn (London: Science Paperbacks)

Bradley J V 1968 *Distribution Free Statistical Tests* (New York: Prentice-Hall)

Butcher J C and Messel H 1960 *Nucl. Phys.* **20** 15–128

Byckling E and Kajantie K 1973 *Particle Kinematics* (London: Wiley)

Carter L L and Cashwell E D 1975 *Particle Transport Simulation with the Monte Carlo Method* (Oak Ridge, TN: USERDA Technical Information Center)

Chandrasekhar S 1943 *Rev. Mod. Phys.* **15** 1–89

Fano U, Spencer L V and Berger M J 1959 *Handbuch der Physik* ed. S Flugge **38**(2) 660

Fishman G S 1978 *Principles of Discrete Event Simulation* (New York: Wiley-Interscience)
Hammersley J M and Handscomb D C 1964 *Monte Carlo Methods* (London: Chapman and Hall)
Heitler W 1954 *The Quantum Theory of Radiation* (Oxford: Clarendon)
Kahan B C 1961 *J. R. Stat. Soc.* (A) **124** 227–39
Knuth D E 1981 *The Art of Computer Programming* vol. 2 2nd edn (Reading, MA: Addison-Wesley)
Metropolis N, Rosenbluth A W, Rosenbluth M N, Teller A H and Teller E 1953 *J. Chem. Phys.* **21** 1087–92
Myhre J M 1970 *Markov Chains and Monte Carlo Calculations in Polymer Science* ed. G G Lowry (New York: Marcel Dekker)
Reichl L E 1980 *A Modern Course in Statistical Physics* (London: Edward Arnold)
Ripley B D 1983 *Proc. R. Soc.* A **389** 197–204
The Rand Corporation 1955 *A Million Random Digits with 100,000 Normal Deviates* (Glencoe, IL: Free Press)
Wilson R W 1952 *Phys. Rev.* **86** 261–9

Chapter 8

Applications of the Monte Carlo Method

8.1 Introduction

Monte Carlo methods have been used in physics in a wide variety of topics, ranging from impurity diffusion in solids to the evolution of galaxies. However, it is in the fields of thermal physics and reactor physics that they have been found of greatest value. We devote most of our description to examples from these two fields, starting with the latter topic. In the last section we explain the use of Monte Carlo methods in calculations related to quantum field theory. It should be realised that the merit of the Monte Carlo method is seen when it is applied to problems of great complexity. For pedagogic reasons it will sometimes be necessary to take examples shorn of much of their complex details; in these cases, use of the method serves more as an illustration of the approach which should be taken to the problem than as an optimum way of solving it.

8.2 The Reactor Criticality Problem

In the design of a nuclear reactor computer methods find very wide application, from the use of finite difference and finite element methods in the structural and heat flow aspects to Monte Carlo methods in the estimation of power output. From the point of view of the latter calculations, the particles of greatest interest are the neutrons, whose diffusion in the core and shield presents a classic example of a stochastic process. Two aspects of the neutron gas are important from the point of view of design, the development with time of the neutron density in the assembly and the containment of these neutrons to shield the exterior region from their harmful effects. This latter particle transport problem does not differ significantly from that discussed by way of an example in the previous chapter. Here, the application of Monte Carlo methods to the former, the so called 'criticality' problem, will be discussed.

As far as neutrons are concerned (Henry 1975) a reactor is composed of three types of substances, absorbing, moderating, i.e. causing the neutrons to slow down, and fissionable material. The distribution of neutrons in the reactor at any time, which initially might just be a single particle from a spontaneous fission, is characterised by a directional density distribution $\rho(\mathbf{r}, \boldsymbol{\omega}, E, t)$. This means that in a volume element $\mathrm{d}V$ at the location $\mathbf{r}$, the number of neutrons having an energy in the range $(E, E+\mathrm{d}E)$, and travelling in an element of solid angle $\mathrm{d}\Omega$ about the direction of the unit vector $\boldsymbol{\omega}$, which has components $(\sin\theta\cos\phi, \sin\theta\sin\phi, \cos\theta)$, is $\rho(\mathbf{r}, \boldsymbol{\omega}, E, t)\,\mathrm{d}V\,\mathrm{d}\Omega\,\mathrm{d}E$. The competition between absorption, escape from the assembly and regeneration of neutrons in fission reactions causes this distribution to evolve with time. Whatever the intitial distribution, we will see that, in general, the density of neutrons throughout the system will either decay away or increase uncontrollably, depending on the size and composition of the assembly. Intermediate between these two cases a unique arrangement exists (in practice achieved by absorbing control rods) where, apart from statistical fluctuations, the distribution settles down to a steady density, i.e. the rate at which neutrons are produced in fissions just balances the rate at which they escape or are absorbed. This is the critical state, and the power of the reactor is determined by the rate at which fission occurs in this state. Obviously, knowledge of the fuel–moderator–absorber arrangement which corresponds to the critical state, and of the spatial density of neutrons, is vital for the design of a reactor; for illustration purposes we will just confine ourselves to the establishment of criticality. This, we will see, can be characterised by a single parameter λ_0, the *time constant*.

Writing $\Psi(\mathbf{r}, \boldsymbol{\omega}, E, t) = v\rho(\mathbf{r}, \boldsymbol{\omega}, E, t)$, where v is the speed of the neutron, the time development of the function Ψ is determined by the Boltzmann transport equation (for details see §8.2 of Henry 1975) and this can be written in the form

$$\frac{\partial \Psi}{\partial t} = L\Psi. \tag{8.1}$$

Here, L is a complicated integro-differential linear operator which, assuming external neutron sources and spontaneous fissions are absent, accounts for neutron absorption, scattering and induced fission; its detailed form need not concern us at present. If solutions to (8.1) of the form $\Psi_i(\mathbf{r}, \boldsymbol{\omega}, E)\,\mathrm{e}^{\lambda_i t}$ are assumed to exist, the function $\Psi_i(\mathbf{r}, \boldsymbol{\omega}, E)$ must satisfy eigenvalue equations of the form

$$L\Psi_i = \lambda_i \Psi_i \tag{8.2}$$

and the general solution to (8.1) may be written

$$\Psi(\mathbf{r}, \boldsymbol{\omega}, E, t) = \sum_i C_i \Psi_i(\mathbf{r}, \boldsymbol{\omega}, E)\,\mathrm{e}^{\lambda_i t} \tag{8.3}$$

with the coefficients C_i determined by the initial conditions. For sufficiently large values of t the solution (8.3) will be dominated by the term $\Psi_0(\mathbf{r}, \boldsymbol{\omega}, E)\,\mathrm{e}^{\lambda_0 t}$, where λ_0 is the largest of the eigenvalues of (8.2) (strictly, the eigenvalue with the largest real part which need not be the one of largest modulus). Depending on whether λ_0 is positive, negative or zero, the assembly is said to be supercritical, subcritical or critical.

A Monte Carlo method immediately suggests itself for finding this maximum eigenvalue. Because the eventual behaviour in time is independent of the intial neutron distribution, starting with an arbitrary spatial and energy distribution in an assembly of specified size and composition and simulating the diffusion of these neutrons and their successors, including all relevant physical processes, the time evolution may be mapped out. As time progresses the spatial and energy distribution will tend to a limiting value $\Psi_0(\boldsymbol{r}, \boldsymbol{\omega}, E)$; when this steady state has been reached, the uniform multiplication rate may be calculated from the numbers of neutrons, or more generally their weights w_i, at two separated times t and $(t + \tau)$,

$$\lambda_0 \simeq \frac{1}{\tau} \ln \left[\sum_i w_i(t+\tau) \left(\sum_i w_i(t) \right)^{-1} \right].$$

The use of this method to estimate λ_0 is described in Davis (1963).

Analogue simulation of a system in time, however, is not the best way to estimate criticality, in part because of the different roles played by the collisions of neutrons of different velocities in the process of settling down to the asymptotic distribution. It can be argued (see, for example, §§31-14 and 32-16 of Fox 1962) that the differentiation of neutrons by *generation* yields a superior parameter, the *multiplication constant*, denoted by k. This is defined as the ratio of the average number of neutrons in a generation to the number in the previous generation, and in the equilibrium state it assumes the value unity, i.e. k is a function of λ_0 and $k(0) = 1$. Here, by first generation neutrons we mean neutrons arising from fissions induced by neutrons of the initial distribution, second generation are those produced by fissions induced by first generation neutrons, etc. Although the neutron population at any instant does not correspond to a unique generation, because of the different particle velocities and the stochastic nature of the collisions, the time constant λ_0 vanishes when the corresponding multiplication constant is unity. This means that generation number, rather than time, can be taken as the parameter of evolution. By starting with a fixed number of neutrons, having an arbitrary spatial and energy distribution in the initial (zeroth) generation and simulating their passage through the assembly, first generation neutrons having a different spatial and energy distribution will be generated. If we continue to simulate the system in this way, these distributions will eventually settle to a steady state, as will the ratio $k_i = n_{i+1}/n_i$, where n_i is the total number of neutrons in the ith generation.

For s successive generations, starting with the jth generation after this settling down has taken place, a weighted estimate of the multiplication constant (§8.7 of Hammersley and Handscomb 1964) can be formed, i.e.

the cumulative estimate

$$k = \sum_{i=j}^{j+s-1} n_i k_i \left(\sum_{i=j}^{j+s-1} n_i \right)^{-1} = \frac{N - n_j}{N - n_{j+s}}. \tag{8.4}$$

Here, $N = n_j + n_{j+1} + \ldots n_{j+s}$.

PROJECT 8A: ENRICHMENT OF NATURAL URANIUM

The energy of neutrons produced in fission is typically 10^6 eV. However, the cross section for fission is greatest when the neutron energy corresponds to thermal energies, i.e. is of the order of 1/40 eV, so reactors usually employ moderating material in which the neutrons are slowed down through multiple collisions. In graphite, typically 100 scatterings are required for thermalisation. Monte Carlo methods can be used in this case (see, for example Mendelson 1968). However, to avoid the very great amount of tracking involved, as well as the complexity of target motion which may not be ignored at thermal energies, we will consider, as an example, the less important case of a fast assembly containing only uranium.

The average numbers of neutrons emitted in the fast fission of ^{235}U and ^{238}U are 2.50 and 2.66 respectively. They have a spectrum of energies which extends beyond 10 MeV, with an average value of approximately 2 MeV. The cross sections for absorption and fission, σ_a and σ_f, of the two isotopes depend on energy. They are given in table 8.1, averaged over the energy spectrum, where the values are in barns (1 barn $= 10^{-28}$ m^2). It is really the energy spectrum in the equilibrium state which should be used for this averaging procedure and this could be found by incorporating the energy dependence of the cross sections into the simulation, suggested for a later refinement of the project.

Allowing for escape, no mass of natural uranium, for which the enrichment factor $f = m_{235}/(m_{235} + m_{238}) = 0.0072$, where m_{235} is the mass of ^{235}U, attains criticality. In this project we seek to determine the enrichment factor at which a sphere of uranium of radius 10 cm will become critical by estimating the multiplication constant, k, for different compositions. Successive stages of complexity approaching the real situation can be built into the project. Initially, all neutrons should be considered mono-energetic, i.e. the cross sections in the table always apply, and neutron scattering may be ignored; the only processes are escape, absorption and fission. The number

Table 8.1 Energy-weighted neutron cross sections.

	^{235}U	^{238}U
σ_a	0.108	0.08
σ_f	1.25	0.30

of neutrons, ν, emitted in a fission has the probability distribution $p(\nu)$ shown in table 8.2, and they can be assumed to be emitted isotropically. Let us take a density corresponding to $N = 4.810^{22}$ uranium atoms/cm^3.

Table 8.2 Fission neutron multiplicity distribution.

ν	0	1	2	3	4	5	6
$p(\nu)$	0.027	0.158	0.339	0.302	0.130	0.034	0.010

The number of neutrons in the zeroth generation must be chosen, say 50, as well as their spatial distribution. For this, two possibilities suggest themselves; that they all start at the centre of the sphere or that they have their origin randomly, but uniformly, distributed throughout the sphere. Intuitively, the latter might be expected to lead to earlier convergence to the equilibrium distribution. However, the greater probability of early escape in many cases may cancel this out.

In the course of tracking the neutrons from generation to generation two questions must be answered. (i) How do we know when the equilibrium distribution has been attained? (ii) When it has been attained, how can we efficiently estimate the degree of criticality, i.e. the multiplication factor k? For the latter, the cumulative estimator (8.4) is suggested, for the former some trial and error will be required, e.g. look for a settling down in the behaviour of the multiplication factor, k_i, from generation to generation and in the spatial moments $\langle r \rangle$, $\langle r^2 \rangle$ etc of the distribution of fission points.

Each neutron in a given generation in the simulation must have the following parameters associated with it; its radial distance r_i, the angle, θ_i, its velocity makes with the radial direction, or its cosine, μ_i (if scattering is included a more complete specification of the velocity vector will be required). See, for example, Davis (1963). For later modification it will be helpful to also provide for a weight w_i for each particle, at present equal to unity, and an energy E_i. Starting with n_0 (of the order of 50) particles in the zeroth generation, their distances to interaction (or escape if it would occur outside the sphere) must be determined in turn from the exponential distribution $\exp(-s\lambda^{-1})\lambda^{-1}$ where $\lambda = 1/\Sigma\, N_i \sigma_T^{(i)}$ where N_i, the number density of atoms of isotope i, depends on the enrichment factor f, and $\sigma_T^{(i)}$ is the total cross section for that isotope. The nature of the collision, absorption or fission (or scattering if included) can be determined on the basis of the cross sections in table 8.1. If fission occurs, all the neutrons emerging, selected using equation (7.7) on the basis of the distribution in table 8.2, must be labelled and stored for processing in the next generation. Note that isotropy of emission corresponds to $p(\mu)\,\mathrm{d}\mu = \mathrm{d}\mu/2$, where $-1 \leqslant \mu \leqslant 1$ ($\mu = \cos\theta$, where θ is the polar angle of an emitted neutron), while the radial coordinate for the newly born neutrons is $(r_i^2 + s^2 + 2\mu r_i s)^{1/2}$, where r_i was the originating point of the neutron which produced the fission. Before it becomes possible to establish the equilibrium distribution it may be that the number of neutrons in a generation has grown to an unmanageably large number, or has dwindled to very few. In such a case the technique of Russian roulette, or splitting, described in §7.6 may be employed. In practice, it will be helpful to renormalise the number of neutrons in each generation to the number in the initial sample. If, for example, this was 50, and the $(i+1)$th generation consisted of 107 particles, only 50 of these (chosen at random) would be processed, but each with its weight multiplied by 107/50, i.e. $w_{i+1} = 107w_i/50$. The quantities n_i in (8.4) would then be replaced by $50w_i$. Other methods for controlling the number of particles may be devised, for which an interactive facility in the computing will prove very helpful, as it will in the early stages of the calculation when an idea of the magnitude of the enrichment factor is being sought. By plotting the multiplication constant versus the enrichment factor, the value of the latter corresponding to criticality may be interpolated.

Because of energy loss due to inelastic scattering and variable probabilities of escape, the energy spectrum of neutrons in the equilibrium state will not be the same as the production spectrum; the important ratio of absorption to fission cross section will then be different from that in table 8.1. To perform this calculation more accurately the program would need to be expanded to take into account the variation of the various processes with neutron energy. To a first approximation (see also Exercise 8.1), the energy distribution of fission neutrons corresponds to a Maxwell–Boltzmann distribution for their velocity (see Exercise 7.12). Specifically, $E_n = 1.29u$ MeV, where u is distributed as

$$p(u)\,\mathrm{d}u = \frac{2}{\sqrt{\pi}}\,u^{1/2}\,\mathrm{e}^{-u}\,\mathrm{d}u. \tag{8.5}$$

The formulation for elastic and inelastic scattering may be found in Davis (1963).

8.3 Thermal Systems in General

The study of thermal systems has for a long time been formulated in terms of statistical concepts, and it is not surprising therefore that Monte Carlo methods should find applications in this field. These methods are useful for two reasons; their intrinsic affinity with the phenomena under study and their ability to extract useful information from a theoretical formalism (e.g. descriptive of a many-interacting particle system) which can be interpreted otherwise only in the limit of severely restrictive approximations. The methods of calculation are, analogously, of two types; (i) direct simulation of the time evolution of a system, e.g. tracking the motion of its constituents, which gives us an image of its real time development, and (ii) integration of very complicated expressions which summarise the underlying stochastic theory. The former method is especially important for studies of fluctuations, or studies of systems not near equilibrium. In the latter context, it should be observed that round-off errors will unavoidably introduce some irreversibility into the results, which must be distinguished from any physical irreversibility in the system being simulated. The systems we study may be isolated or in thermal contact with the environment. The type of problem not easily amenable to analytical methods might concern the occurrence of a phase change in a system of molecules interacting via a specified potential, or the heat capacity of a system of dipoles interacting via a quantum exchange force. Indeed, like thermodynamics itself, the methods developed in such problems are found to have much wider applicability than was suggested at the time of their introduction and they are now of great importance in Monte Carlo calculations in quantum field theory. We will start our discussion with a closed system.

PROJECT 8B: AN IDEAL GAS

This simple example illustrates how the fundamental principle of statistical mechanics, namely that all microscopic configurations consistent with the constraints on a system are a priori equally probable, can be used to investigate the properties of an isolated ideal gas. In particular, it is found that the velocity distribution of the molecules is of the Maxwell–Boltzmann form when the number of molecules is great.

The set of all the space, spin, and momentum coordinates of the molecules in the system at one instant defines a single point in a multidimensional phase space, and each such point defines a unique microstate of the system. Possible configurations will correspond to certain permitted regions in this space, which, by the fundamental principle, will be uniformly densely populated with such points. For an isolated system of ν non-interacting molecules, of mass m, the energy is given in terms of the momentum components p_i by

$$U = \sum_{i=1}^{\nu} |\boldsymbol{p}_i^2|(2m)^{-1} = \sum_{i=1}^{N} p_i^2(2m)^{-1}$$

where $N = 3\nu$. Therefore, since the energy is fixed, all points in the permitted region of the phase space must lie on the surface of a hypersphere of radius $p = \sqrt{2mU}$ in the momentum subspace. Representative microstates of the gas can thus be obtained by choosing points uniformly at random on the surface of this sphere, which is embedded in the N-dimensional space. By selecting a sequence of such states, the average behaviour, e.g. the distribution in molecular velocities, of an ensemble of such systems can be studied.

We now indicate how such states may be sampled. $(N-1)$ angles are required to specify a point on the sphere. For example, with $N = 3$ the probability distribution of a point is $d\Omega/4\pi$, or, writing $\boldsymbol{\theta} = (\theta, \phi)$,

$$\frac{d\Omega}{4\pi} = \left(\frac{\sin\theta}{2}\right) d\theta \frac{d\phi}{2\pi} = p(\theta)\, d\theta\, p(\phi)\, d\phi.$$

It can be shown that, writing the angular coordinates of a point on the N-dimensional sphere $\boldsymbol{\theta} = (\theta_1 \ldots\ldots \theta_{N-1})$, we have the separable form

$$p(\boldsymbol{\theta})\, d\boldsymbol{\theta} = \prod_{i=1}^{N-1} p_i(\theta_i)\, d\theta_i$$

where

$$p(\theta_1)\, d\theta_1 = d\theta_1/2\pi \qquad 0 \leqslant \theta_1 \leqslant 2\pi$$

$$p_i(\theta_i)\, d\theta_i = K_i \sin^{i-1}\theta_i\, d\theta_i \qquad 0 \leqslant \theta_i \leqslant \pi \quad 2 \leqslant i \leqslant (N-1)$$

and the normalising constants, K_i, are given by

$$K_i = \Gamma[(i+1)/2][\sqrt{\pi}\,\Gamma(i/2)]^{-1}.$$

In this equation Γ denotes the gamma function. The coordinates of the point are then given by

$$p_1 = p \cos\theta_{N-1}$$

$$p_i = p \sin\theta_{N-1} \sin\theta_{N-2} \ldots \sin\theta_{N-i+1} \cos\theta_{N-i} \qquad 2 \leqslant i \leqslant (N-1)$$

$$p_N = p \sin\theta_{N-1} \sin\theta_{N-2} \ldots \sin\theta_2 \sin\theta_1.$$

Writing $z = \cos\theta$, the distribution $p_i(\theta_i)$ can be cast in the form

$$p_i(z_i) = K_i(1 - z_i^2)^{(i/2)-1}\, dz_i$$

where $-1 \leqslant z_i \leqslant 1$ and $2 \leqslant i \leqslant (N-1)$. This is just the distribution (7.13) previously discussed in an illustration of importance sampling, which can thus be used for $i \geqslant 2$.

For small samples, say less than 20 molecules, compare the average distribution of velocity, and of one component, say v_x, with the Maxwell–Boltzmann expressions and find the behaviour of $\overline{v^2}/(\bar{v}_{\mathrm{MB}})^2$ versus N, where $(\bar{v}_{\mathrm{MB}})^2 = 2U/Nm$. An alternative formula for selecting the velocity components can be found in Exercise 8.2, while further information can be found in Sauer (1981).

8.4 Interacting Thermal Systems

The previous simple example was rather artificial in the sense that such isolated systems are seldom of interest in physics; we are more likely to be concerned with a system interacting with its environment and we shall restrict ourselves to cases where that interaction is one of equilibrium, i.e. it can be characterised by a temperature T. Furthermore, we usually wish to describe the state of the system in terms of macroscopic, i.e. thermodynamic, variables rather than microscopic ones. In terms of such variables, of course, not all values are equally probable, rather the equiprobable microstates of the system plus its environment result in a probability distribution for the microstates of the system related to its total energy, as given by the Boltzmann distribution. It is on the basis of this distribution that groupings of configurations of the system can be investigated. Finally, an obvious limitation on this kind of simulation is the small number of molecules for which, because of computer time, calculation is possible. When surface or interface effects are not of interest, the use of periodic boundary conditions, explained below, enable the system to be viewed as being immersed in an infinite sample, in which case the results may be more representative of a much larger system.

Each possible set of coordinates of the particles of the system (e.g. position, momentum, spin etc), denoted by $\boldsymbol{\alpha}'$, defines a configuration of the system and at equilibrium the fundamental probability distribution of these microscopic coordinates is given by the Boltzmann distribution,

$$p(\boldsymbol{\alpha}',T)\,\mathrm{d}\boldsymbol{\alpha}' = \exp[-\beta E(\boldsymbol{\alpha}')](Z')^{-1}\,\mathrm{d}\boldsymbol{\alpha}' \tag{8.6}$$

where $\beta = 1/k_{\mathrm{B}}T$ and $E = \Sigma\, p_i^2/2m_i + \Phi(\boldsymbol{\alpha}')$—the total energy (more generally the Hamiltonian). Z' is a normalising constant which is known as the partition function and is given by

$$Z' = \int \mathrm{d}\boldsymbol{\alpha}' \exp[-\beta E(\boldsymbol{\alpha}')].$$

In cases where the force between particles does not depend on their velocities, integration over the momentum coordinates can be carried out and, denoting the remaining phase space coordinates by $\boldsymbol{\alpha}$, one obtains

$$p(\boldsymbol{\alpha}, T)\,\mathrm{d}\boldsymbol{\alpha} = \exp[-\beta\Phi(\boldsymbol{\alpha})]Z^{-1}\,\mathrm{d}\boldsymbol{\alpha}. \tag{8.7}$$

Here, Z is the configurational partition function, e.g. for a system of N molecules it would have the form:

$$Z(N, B) = \int \mathrm{d}^3x_1 \ldots\ldots \mathrm{d}^3x_N \exp[-\beta\Phi(x_1, \ldots, x_N)]. \tag{8.8}$$

Hence, at equilibrium at T ($T = 1/k_{\mathrm{B}}\beta$) the mean value of any thermodynamic quantity, e.g. the internal energy, denoted generally by F, is

$$\langle F(T)\rangle = \int F(\boldsymbol{\alpha})p(\boldsymbol{\alpha}, T)\,\mathrm{d}\boldsymbol{\alpha}. \tag{8.9}$$

The extent of fluctuations about this average may also, in some cases, be related to macroscopic thermodynamic quantities. For example, in the case of a component of the internal energy, E, it can be readily shown (see §2.5

of Mandl (1971)) that the corresponding contribution to the specific heat for a system of N particles is given by

$$\frac{C}{Nk_B} = \left(\frac{\beta^2}{N}\right)\sigma_E^2 \tag{8.10}$$

where σ_E^2 is the variance on E.

In so far as we know $F(\boldsymbol{\alpha})$ and the probability distribution (8.7), our earlier discussion (§7.6) of importance sampling suggests a very direct way of evaluating the integral (8.9), namely select points from the distribution $p(\boldsymbol{\alpha}, T)$ and average the corresponding values of F. A major problem arises because we don't know $p(\boldsymbol{\alpha}, T)$ explicitly; it involves the normalising constant Z^{-1}, itself a multidimensional (in practice possibly 10^{23}-dimensional!) integral whose evaluation would be a major Monte Carlo problem in itself. On the other hand, evaluating (8.9) by random sampling of the phase space points, like we did in Project 8B, is very inefficient. Most of the points uniformly selected in this space of enormous dimensions would fall in regions which, because of the $e^{-\beta\Phi}$ term, would make negligible contribution to the sum. The method of Markov chains, described in §7.4.2 above, can be used to approximate the method of importance sampling in this case, and we again describe this method in the context of a concrete example below. It is recommended that Exercise 8.3 be done before continuing with Project 8C.

PROJECT 8C: A MAGNETIC DIPOLE SYSTEM—THE ISING MODEL

It would seem logical to illustrate the above discussion by considering the thermal properties of a system of gas molecules interacting through a separation-dependent potential, $\boldsymbol{\alpha} = \{x_i\}$, and indeed it was in the study of such systems that the methods were developed. For introductory purposes it will be simpler to consider a system with a finite number of discrete possible configurations. This will introduce all the techniques relevant to the somewhat more complex gas/liquid simulation. The system we describe consists of N magnetic dipoles and the kind of model used is known as the *Ising model*. The N magnetic dipoles, e.g. electrons, are arranged at fixed points on a lattice. With a uniform magnetic field, B, applied the potential energy of the system is given classically by

$$\Phi = -\sum_{i=1}^{N} \boldsymbol{\mu}_i \cdot \boldsymbol{B} + \sum_{i \neq j} f(r_{ij})\boldsymbol{\mu}_i \cdot \boldsymbol{\mu}_j \tag{8.11}$$

where r_{ij} denotes the separation between dipoles i and j. Quantum mechanically, $\boldsymbol{\mu}_i = g\mu\boldsymbol{s}_i$, where $\boldsymbol{s}$ denotes the spin vector, and different possibilities for the dipole–dipole interaction exist. In the Ising model, only pairs (i, j) which are nearest neighbours on the lattice are included and the coefficients f_{ij} are taken as constant. Taking the $\boldsymbol{B}$ field as the reference direction, on which the spins ($s = 1/2$) have only two projections, we define a two-valued spin coordinate $\alpha_i = \pm 1$ (i.e. $\mu_i = \text{constant} \times \alpha_i$). A configuration, $\boldsymbol{\alpha}$, of the system will be specified in terms of these coordinates and we rewrite (8.11), remembering that we are only dealing with the magnetic aspect of the system. Contributions from lattice vibrations etc are considered separately if necessary:

$$\Phi(\boldsymbol{\alpha}) = -H \sum_{i=1}^{N} \alpha_i - J \sum{}' \alpha_i \alpha_j$$

where Σ' indicates summation over all pairs of sites which are nearest neighbours and H and J are constants. This can be rewritten as

$$\Phi(\boldsymbol{\alpha}) = -H \sum_{i=1}^{N} \alpha_i - \frac{J}{2} \sum_{i=1}^{N} \alpha_i \sum_{j(i)} \alpha_j \qquad (8.12)$$

where the summation over j only runs over those sites which are nearest neighbours of the site i and the factor of 1/2 arises from the double counting of each (i, j) pair in the sum. Here, $\boldsymbol{\alpha}$ denotes a given configuration in which n_+ of the dipoles have $\alpha_i = +1$ and $(N - n_+)$ have $\alpha_i = -1$. In zero applied field we have $H = 0$; the constant J may be positive or negative reflecting situations where the energy will be least for parallel or antiparallel alignments of the dipoles. This provides a simple model for ferromagnetic and antiferromagnetic substances respectively. The analytical treatment of even a simple system of this kind is formidable and has only been solved for one and two-dimensional cases, and then only for infinite systems. The kind of quantities which are of interest for comparison with experiment are the energy as a function of temperature, and hence the specific heat, the magnetisation or 'long-range order' of the system, etc. For example, the magnetisation per dipole of any configuration is $M^{(\alpha)}$ where $M^{(\alpha)} = (1/N) \Sigma \alpha_i$, and the corresponding mean magnetisation at T is given by the average over configurations,

$$\langle M(T) \rangle = \sum_{\alpha} M^{(\alpha)} p(\boldsymbol{\alpha}, T) \qquad (8.13)$$

with the probability distribution given by (8.7). Ehrman *et al* (1968) and Landau (1976) consider this problem.

The selection of configurations $\boldsymbol{\alpha}$ is done using Markov chains, the principles of which were described in §7.4, where it is seen that the formula is not unique. If the system consists of a total of N dipoles, states distributed according to (8.7) are selected by carrying out a random walk from state to state; at each step, non-zero transition probabilities only exist for those N (out of a total of 2^N) states which differ from it by the alignment of a dipole at *one* of the N sites. The selection procedure is as follows.

(i) Start with an arbitrary configuration $\boldsymbol{\alpha}_i$.
(ii) Choose with equal probability, equal to $1/N$, a tentative configuration

which is one of the N distinct configurations obtainable by reversing a spin at one site only and call it $\boldsymbol{\alpha}_j$; calculate, using (8.12), $\Delta\Phi$ corresponding to the reversal of the spin there.

(iii) If $\Delta\Phi < 0$, reverse this spin to get a new configuration, $\boldsymbol{\alpha}' = \boldsymbol{\alpha}_j$ (this corresponds to $p(\boldsymbol{\alpha}_j) > p(\boldsymbol{\alpha}_i)$ and $W_{ij} = 1$ in (7.14)).

(iv) If $\Delta\Phi > 0$, reverse the spin if $e^{-\beta\Delta\Phi} > \xi$, where ξ is a uniform random deviate on (0, 1), to get $\boldsymbol{\alpha}' = \boldsymbol{\alpha}_j$. Otherwise, keep the old configuration $\boldsymbol{\alpha}' = \boldsymbol{\alpha}_i$ (corresponding to $W_{ij} = \exp[-\beta\Phi(\boldsymbol{\alpha}_j)]/\exp[-\beta\Phi(\boldsymbol{\alpha}_i)]$ in (7.14)).

It has been argued above that these transition probabilities eventually lead to a succession of configurations, which are characteristic of the Boltzmann distribution. At this stage, quantities like (8.13) can be evaluated by averaging over later steps in the chain.

While there may be times when very small systems are of interest, e.g. inclusions of a magnetic substance in a background material (in which case surface effects would be of intrinsic interest), it is usual for comparisons to be made with results on macroscopic systems. Remembering that there are 2^N distinct configurations of such an assembly, it is obvious that the size of the system for which such a relatively complicated algorithm can be practically employed will not anywhere approach a macroscopic one. To approximate a macroscopic system it is usual to consider a system with periodic boundary conditions, e.g. in a cubic array the basic cell of N dipoles is visualised as being surrounded throughout the calculation by replicas of itself on all sides. Because the present model only involves nearest neighbour interactions, this encirclement can be reduced to single layers of the replicas, including the interaction of these dipoles with their neighbours inside the cell.

A series of increasingly sophisticated programs can be developed by stages to illustrate these ideas. Starting with a cubic system of dipoles, e.g. $4 \times 4 \times 4$, in an external magnetic field with no dipole–dipole interaction, i.e. $H = 1$, $J = 0$ in (8.12), investigate the magnetisation and specific heat as a function of β in the region $0 < \beta \lesssim 0.15$, and also the magnetic entropy, and compare them with known results, e.g. in §§3.1 and 3.2 of Mandl (1971). The initial values of the coordinates at each lattice site, $\alpha_i = \pm 1$, can be assigned in either of two ways; the values $+1$ or -1 assigned at random to the various sites (a 'hot' start), or all initial values are the same, e.g. $\alpha_i = +1$ (ordered or 'cold' start). In the successive sampling of configurations, using the procedure described above, one can choose successive points at random throughout the lattice for the next tentative spin flip, or you can move systematically from point to neighbouring point. The latter may be a bit easier in computing terms. In finding the average of any quantity, e.g. the energy, over the random walk among configurations there will be the problem of deciding when convergence to the Boltzmann distribution is achieved (remember, §7.4, that it occurs 'eventually'), i.e. when is the influence of the initial arbitrary configuration negligible? This can be done by trial and error. If a more objective criterion is desired, running two parallel walks, one from a cold start and the other from a hot start, and seeing when their averages begin to agree will serve as a test. Only after this stage can a calculation of the ensemble average start, and even then some care must be taken because, although the states accessed in the sampling are now representative of the distribution of states of the system, they are strongly correlated. Each successive phase space point differs from the previous one by, at most, a change in one coordinate. Accordingly, all configurations should not be included, e.g. only every 64th configuration might be included in finding the averages. More details on treating the statistics of the results can be found in Wood (1968) and in §A.2 of the Appendix. Note that for small systems the calculated entropy will differ from theoretical values because of the influence of periodic boundary conditions. Their effect in this case can be investigated by varying the size of the system.

To extend the program to investigate Ising-type models, e.g. in zero magnetic field, $H = 0$, $J = 1$ in (8.12), all that will be required is to revise the subroutine for calculating $\Delta\Phi$ in the procedure. These models are important because it is known that, in two dimensions for an infinite system, they can give rise to a phase transition. At a unique temperature the specific heat becomes singular, and below this temperature there is a spontaneous long-range order (magnetisation in the case of a ferromagnet). Because the analytical theory of such cooperative phenomena, accompanied as they are by infinite fluctuations, is especially difficult, much work has gone into Monte Carlo simulations along these lines (Ehrman *et al* 1968, Landau 1976). For several values in the region $0.1 < \beta < 0.5$, the root mean square magnetisation $\sqrt{\langle M^2 \rangle}$, averaged over many configurations, and β^2 times the variance in the energy (which equals $\langle E^2 \rangle - \langle E \rangle^2$, cf (8.10)) over these same configurations should be determined and plotted against β to search for evidence of a phase transition in such a small system. A further refinement would be to introduce periodic boundary conditions around the cube and see their influence.

In 'mathematical experiments' like that described above it is not always easy to write down in advance a precise strategy, some development of the method, very much like the setting up of a physical experiment, is an intrinsic part of the operation, the following suggestions may be helpful in the present cases.

In using (8.12) to calculate $\Delta\Phi$, because only nearest-neighbour pairs are involved, only a small number of quantities need be calculated. For the case of a simple cubic lattice, if the coordinate α_k is reversed we find

$$\Delta\Phi = 2H\alpha_k + 2J\alpha_k \sum_{j(k)} \alpha_j.$$

For serious work it will be much faster if the coordinates ± 1 are represented by single bits (0 and 1) of binary words in the computer, and the walk executed by operating on these words.

If plenty of time is available, an alternative to employing periodic boundary conditions is to investigate systems on different size lattices and try to find some scaling laws by which an extrapolation can be made to asymptotic behaviour.

8.5 Random Walks and Macromolecular Chains

Matter in a condensed form can exhibit very different physical properties; for example, the thermal coefficient of expansion of most substances is positive while for rubber-like solids we can find negative values. Also, there is an obvious qualitative difference in the elasticity of such substances when compared to that of normal solids. Analytical methods can be applied to normal solids with great success, especially when they are in crystalline form, while the absence of rubber-like solids (notwithstanding their great practical importance) from most undergraduate texts attests to the greater difficulty of describing macromolecular substances by standard analytical methods. This is a field where computational methods, and in particular the Monte Carlo method, have been notably successful and although exact results in a few cases have been obtained by analytical methods in recent years, numerical methods still have much to offer.

A polymer molecule is a chain of repeating smaller molecular units, a total of $(N+1)$ subunits in all. For a specified number N of interunit bonds there are a great number of spatial configurations of the molecule, arising from oscillations and rotations of segments about the bonds. These, in turn, are determined by intramolecular forces and, in the polymer immersed in a solute, forces between molecules of the chain and those of the solute. The study of this field, polymer science, is a very large one; we will confine ourselves to some elementary aspects of it.

As an introduction, we will consider a simplified model of a polymer molecule built up of N links each of length h and enquire into the root mean square length of the chain averaged over many of its configurations. If, in the construction, successive molecules can adopt an orientation at random, like a particle undergoing Brownian motion, our problem coincides with the three-dimensional random walk (see Exercise 7.4), analogous to the flight of a seagull discussed many years ago by Rayleigh, and the results are well known (Chandrasekhar 1943). The number of arrangements, which is related to the entropy of the polymer, where the last $(N+1)$th molecule lies at a distance $\boldsymbol{R}$, $\mathrm{d}^3\boldsymbol{R}$ from the first one is

$$p(\boldsymbol{R})\,\mathrm{d}^3\boldsymbol{R} = (b\sqrt{\pi})^{-3}\exp(-\boldsymbol{R}^2/b^2)\,\mathrm{d}x\,\mathrm{d}y\,\mathrm{d}z \tag{8.14}$$

with $b^2 = 2Nh^2/3$ (cf the central limit theorem, §A.2 of the Appendix). The corresponding root mean square end-to-end distance averaged over all configurations is

$$\sqrt{\langle R^2 \rangle} = hN^{1/2} \tag{8.15}$$

a result which holds for such walks in any number of dimensions. This latter quantity, the mean square end-to-end distance $\langle R^2 \rangle$, is characteristic of the dimensions of the chain. However, an alternative parameter, which is more important in some analyses, is the mean square radius of the chain from its centre of mass, that is

$$\langle S_N^2 \rangle = \frac{1}{N+1} \sum_{i=1}^{N+1} \boldsymbol{\rho}_i^2$$

where $\boldsymbol{\rho}_i = \boldsymbol{R}_i - \boldsymbol{R}_{\text{CM}}$, with $\boldsymbol{R}_i$ representing the coordinates of the ith atomic unit, and

$$\boldsymbol{R}_{\text{CM}} = \frac{1}{N+1} \sum_{1}^{N+1} \boldsymbol{R}_i.$$

This can be rewritten as

$$\langle S_N^2 \rangle = \frac{1}{N+1} \sum R_i^2 - R_{\text{CM}}^2 \tag{8.16}$$

(see Exercise 8.4). For the random walk in three dimensions just described, there is a simple relationship between $\langle R_N^2 \rangle$ and $\langle S_N^2 \rangle$, i.e. $\langle R_N^2 \rangle / \langle S_N^2 \rangle = 6$.

There is very limited physical significance in so simple a model for a polymer, but it serves as a starting point for our next example where we will make the model a bit more realistic; henceforth we will set $h = 1$.

PROJECT 8D: COEFFICIENT OF THERMAL EXPANSION OF AN IDEALISED POLYMER

The distribution (8.14) cannot be exact since, physically, the distance R is bounded by $0 \leqslant R \leqslant Nh$. Indeed, at $R = Nh$ there is only one state, and in this region the number of configurations is analogous to the polarisation of a system due to the alignment of individual electric or magnetic dipoles near the saturation limit (cf §3.3 of Mandl (1971)). Just how good equation (8.14) is as a representation of the distribution could be investigated using a Monte Carlo simulation. To increase the contact with reality we add some further constraints; let successive bonds make a unique valence angle θ with respect to their predecessor (refer to figure 8.1). This now resembles actual chains of polyethylene which are arranged with $\theta = \cos^{-1}(1/3)$. Any pair of links, s_i, s_{i+1}, defines a plane and we associate a potential energy of orientation, $\Phi(\phi)$, for the next link with respect to this plane. Physically, this arises mainly from the forces between the atomic groups attached to adjacent segments in the spine of the chain. At any temperature T these forces will

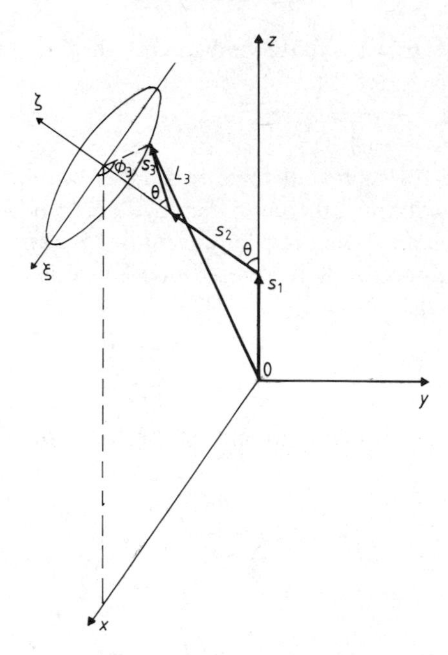

Figure 8.1 Three successive links in a random walk (polymer chain) at each step of which the polar angle θ is fixed while the azimuth ϕ, relative to the ξ axis, which lies in the plane formed by the two previous links, follows a probability distribution. L_3 denotes the length of the three-link segment.

result in an anisotropic distribution relative to the plane because of the Boltzmann factor, this distribution is given by

$$p(\phi)\ \mathrm{d}\phi \propto \exp[-\beta\Phi(\phi)]\ \mathrm{d}\phi \qquad \beta = 1/k_{\mathrm{B}}T.$$

Taking $\Phi(\phi) = V_0(1 - \cos 2\phi)$, where V_0 is a constant, we find

$$p(\phi)\ \mathrm{d}\phi = K \exp(\lambda \cos 2\phi)\ \mathrm{d}\phi \qquad (8.17)$$

where $-\pi \leqslant \phi \leqslant \pi$, $\lambda = \beta V_0$ and K is a normalising constant.

In this project you should investigate, for a fixed number of bonds, N, the dependence of the length of the polymer, L (defined simply as the root mean square end-to-end distance), on the temperature, and from it find the coefficient of expansion $L^{-1}\ \mathrm{d}L/\mathrm{d}T$. Take the chain element length $h = 1$ and a value of $N \gtrsim 30$, depending on how much computer time you have available, and work in terms of $\lambda = V_0/k_{\mathrm{B}}T$, say in the region $0.5 \lesssim \lambda \lesssim 2.0$, i.e. V_0 is positive. This will involve three steps; finding an efficient method for selecting from the distribution (8.17), using this to generate successive steps in the walk and, when sufficient have been generated, evaluating the end-to-end distance L. When the averages over a number of walks at different temperatures have been found, find a suitable way of plotting L against λ which will facilitate taking its derivative.

The first two bonds may be used to define the coordinate axes, $\boldsymbol{s}_1 = (0, 0, 1)$ and $\boldsymbol{s}_2 = (\sin\theta, 0, \cos\theta)$. The coordinates of the end of the ith bond are $\boldsymbol{L}_i = \boldsymbol{L}_{i-1} + \boldsymbol{s}_i$, where the components of the vector $\boldsymbol{s}_i$, in terms of the angle ϕ_i with respect to the plane defined by $\boldsymbol{s}_{i-1}$ and $\boldsymbol{s}_i$, are (see figure 8.1)

$$\boldsymbol{s}_i = \boldsymbol{\xi} \sin\theta \cos\phi_i + \boldsymbol{\eta} \sin\theta \sin\phi_i + \boldsymbol{\zeta} \cos\theta$$

$$\boldsymbol{\xi} = \boldsymbol{s}_{i-1} \operatorname{cosec}\theta - \boldsymbol{s}_{i-2} \operatorname{cotan}\theta$$

$$\boldsymbol{\eta} = (\boldsymbol{s}_{i-2} \times \boldsymbol{s}_{i-1}) \operatorname{cosec}\theta$$

where $\boldsymbol{\zeta} = \boldsymbol{s}_{i-2}$.

*8.6 Polymer Chains with Interactions

A study of the results of the above project would show that, at a fixed λ, the distribution in the end-to-end distance is, as in the unconstrained case, still Gaussian to a very good approximation and the mean square length is also proportional to N, i.e. $\sqrt{\langle R_N^2 \rangle} = fN^{1/2}h$, where $f(\theta, \langle\phi\rangle)$ is a numerical factor. In fact the whole problem could be solved exactly and computational methods are not necessary; however, it serves to lay the groundwork for the real problem, where exact methods are unknown. Even so, the use of such Markov chains to represent polymers in the first approximation can give valuable information for comparison with the physical properties of real polymers.

The Markov model is unrealistic as a representation of a real polymer because at any stage in the chain a new bond direction is selected simply on the basis of the formula, without enquiring what force, which may be attractive or repulsive, will exist between the new segment and existing segments and, in particular, whether the new position chosen to be occupied already contains a molecule (corresponding to the extreme case of repulsion). In other words, we have not taken the previous history of the walk into account. In the case of the flight of the seagull this poses no problem since the same point can be accessed multiply in the walk, but here, obviously, there is a volume surrounding every bond which is forbidden to all later stages in the walk. This is known as the excluded volume effect and the walk is known as a self-avoiding random walk. Physically, of course, the effect arises from interactions between the molecules, at close range the intermolecular force is strongly repulsive. This is not the only component in the interaction however; there will be forces of longer range, both attractive and repulsive, which will play a role in determining the configuration of the molecule (analogous to the interdipole forces in our discussion of the Ising model). These, to a first approximation, might be taken into account by considering the number of non-bonded, nearest-neighbour polymer molecules in the chain. In our first investigation of the self-avoiding walk we shall ignore all such forces except that preventing units from coinciding in space. The evolution of a molecular chain can still be thought of as a random walk, but it is non-Markovian, i.e. the transition probabilities at any stage depend on the previous history of that particular walk. As a consequence, the mathematical properties are profoundly different from an ordinary random walk. The establishment of even simple properties for it, like how many walks of N steps on a simple cubic lattice exist (compare with q^N for a Markovian walk, where q, the coordination number for the walk, is the number of possible choices at each step, six on a simple cubic lattice), cannot be found analytically. Note that a random walk without self-reversal, i.e. the transition probability is zero in the direction from which the walk has most recently come, is still Markovian and, in particular, the relations (8.14) and (8.15) still hold, with different values for some constants. From the practical point of view, the generation of a walk with a continuous distribution of relative orientations and taking previous occupancy into account would seem a hopeless task. Generation will be greatly simplified if we confine the walk to discrete locations on a lattice and in this, fortunately, we are aided by the geometry of real molecular bonds. The most important bond is the C–C, or diamond, bond which orientates in a tetrahedral fashion, and a tetrahedral lattice is a good approximation to some real macromolecular chains. Other manageable arrays in three dimensions are face-centred cubic and simple cubic; for simplicity we will present our discussion in the context of the latter.

We briefly summarise, for self-avoiding walks of N steps, the limited theoretical results so far found; the number of distinct walks is $(q')^N N^\alpha$, where q' is an effective coordination number ($q' \simeq 0.94$ in three dimensions on a simple cubic lattice, $\alpha = 1/6$). The distribution in end-to-end length R is non-Gaussian,

$$p(R) \sim R^2 \exp(-bR^t) \tag{8.18}$$

where $t = 5/2$ in three dimensions. The root mean square length satisfies the relations

$$\sqrt{\langle R_N^2 \rangle} \propto N^{3/4} \qquad \text{in two dimensions}$$

$$\sqrt{\langle R_N^2 \rangle} \propto N^{3/5} \qquad \text{in three dimensions}$$

(compare with $N^{1/2}$ in (8.15) for all dimensions).

We now turn to the determination of these properties by Monte Carlo simulation. This is not straightforward for two reasons; (i) a continuous record of the walk must be maintained throughout its execution and be investigated for every new step to ensure that it is allowed, and (ii) walks are quite likely to come to an impasse, i.e. all tentative steps from a location are forbidden, before the full N steps have been executed. This is known as attrition and obviously will become more severe for large N. The probability when starting a walk that it will equal or exceed N steps without intersecting itself is given (for large N) by

$$P(N) = e^{-N/\nu} \tag{8.19}$$

where ν, the mean number of steps in a walk before it comes to a halt, depends on the dimensionality and a priori transition probabilities of the walk. For a walk on a simple cubic lattice with all 5 choices a priori equal, a value for ν of about 16 is found. The quantity $\lambda \equiv \nu^{-1}$ is known as the attrition constant. Thus the chance on a cubic lattice of completing a 100-link chain is very small. However, typical macromolecules contain $\gtrsim 10^4$ monomers, so we see that even when a program to account for the excluded volume is developed its direct application would lead to a great number of essentially useless walks being generated. This problem has also been encountered in the study of radiation transport (§7.6 above) and we saw how variance reduction, i.e. the pursuit of walks biased to enhance their contribution to the answer, with subsequent removal of the bias, could result in greatly improved efficiency in the calculation.

PROJECT 8E: THE FLORY (Θ) TEMPERATURE FOR A POLYMER

A self-avoiding walk is realised with transition probabilities determined by the intermolecular interactions. Consider walks of N steps ($N + 1$ subunits) on a simple cubic lattice and develop, by stages, a program to determine its properties. As has been pointed out, interactions exist not only between the atoms forming the bonds (strong), but also between any neighbouring atoms in the chain. As a first step, ignore all the interactions except that dictated by the excluded volume effect. In this case there will be no temperature-dependent quantities, but you can investigate $\langle R^2 \rangle$ and $\langle S^2 \rangle$ against N for different N.

Three distinct procedures are involved in such a program; (i) the testing for occupancy of neighbouring sites prior to making the next step of the walk, (ii) the biasing of the walk so as to make use of the many walks which would otherwise abort before N steps, and (iii) the data analysis to extract the required quantities averaged over many walks with the same number of steps. For the first procedure listed the state of each lattice site, i.e. vacant or occupied, must be available and this can be achieved by storing an index in an array with the dimensions of the lattice. The use of conventional numerical arrays to store this information will be very wasteful of memory and, since the specification is very simple, e.g. a zero or one, if you can manipulate binary bits on your computer the assignment of just one binary bit per lattice site will be sufficient. For an exact generation of the walk, at every stage one of the five tentative directions should be chosen with probability 1/5 and if it leads to an occupied site the walk should be abandoned and a new one started. As we pointed out above, this will occur on average after about 16 steps, so to generate a sizable number of chains, even of length 50 steps, some form of biasing will be essential (Wall and Erpenbeck 1959, Mazur and McCrackin 1968). By using the methods of survival biasing and splitting described in the previous chapter, you can make use of the unsuccessful attempts. These should be introduced into your program by stages, using survival biasing for the first attempt, as follows. At the end of the jth bond determine how many of the adjacent sites are vacant, say q_j, and with probability $1/q_j$ choose one of these directions. Provided $q_j > 0$, this ensures continuation of the walk. In the notation of §7.6, the 'absorption' probability is $\mu_{\text{abs}} = 1 - (q_j/5)$, while the total 'interaction' probability is $\mu = 1$. By not allowing 'absorption' to occur, we must assign a weight $1 - (\mu_{\text{abs}}/\mu) = 1 - (q_j/5)$ to the step. If the walk survives N steps it will have a total weight $\Pi_{j=2}^{N}(q_j/5)$, which must be used in evaluating its contribution to any finally calculated quantity. At the expense of increasing complexity this can be extended such that to make a step choice at one point one determines not just the number of potential one-step walks from that point, but also the number of walks of say L steps, and their relative probability, where of course L in practice cannot be very large.

Even with this biased generation, it will still happen that no vacant sites are available at some stage before N steps are completed. As an improvement, the development of your program should introduce the splitting technique, introduced in §7.6, as follows. Divide the generation of the chain into m segments, each of k links long, i.e. $N = mk$. The attrition length, v, plays the role of the absorption mean free path, thus a suitable value for k is 16. After generating the first chain of length k, as described above, replace it with two chains, each with a weight equal to half the weight of the previous chain at the point of splitting. Chains of many different lengths will be generated in this way, although, of course, they will not be independent.

You should compare the dependence of $\langle R^2 \rangle$ on N with that for a Markovian walk and parametrise the distribution in end-to-end distances for a fixed value of N, i.e. estimate the parameter t in the expression (8.18). The important result you should find is that $\langle R^2 \rangle$ increases faster with N than for the Markovian case, i.e. where it is proportional to N (Mazur and McCrackin 1968).

Further refinement for the program will enable you to investigate thermal properties, allowing both for self-avoidance and the remaining interactions between the molecules in the chain. The simplest model (McCrackin *et al* 1973) will be to assume a square-well potential of width equal to the lattice spacing about each molecule so that, apart from the direct bond, only nearest-neighbour unbonded molecules will interact, and with each such interacting pair there is associated an energy ε. For a macromolecule with a fixed number of segments, N, the energy associated with the direct bonds is a fixed quantity and need not be considered. To generate a representative sample of configurations of a chain with N segments, including these intramolecular interactions, we should include Boltzmann factor weightings in the transition probabilities at each step, just as we did in Project 8C, rather than the $1/q_j$ probabilities described above. However, to calculate properties averaged over many chains of N segments, e.g. $\langle R^2 \rangle$, one can generate the chains without including the Boltzmann factor at each step, but keep account of the total number of adjacent, non-bonded, pairs in the chain. One can then weight the properties of the whole chain by an overall Boltzmann factor, which is proportional to its probability of realisation if the correct transition probabilities had been included at every step. If the ith chain (of length N) contains v_i adjacent unbonded pairs, its interaction energy is $v_i\varepsilon$ and its contribution to any result should carry a weight $\exp(-\beta v_i \varepsilon)$, in addition to any weighting introduced due to variance reduction techniques. On the simple cubic lattice the number of nearest-neighbour interactions introduced by adding segment j is $5 - q_{j+1}$ where q_{j+1} is the number of vacant sites at the end of bond $(j+1)$. Thus, the total number of such bonds in the chain is $\Sigma_{j=1}^{N}(5 - q_{j+1})$, or, since $q_2 = 5$, $v_i = \Sigma_{j=2}^{N+1}(5 - q_j)$. The configurational partition function per molecule is

$$Z(T, \varepsilon) = \frac{1}{m}\sum_{i=1}^{m} w_i\, \mathrm{e}^{-v_i\beta\varepsilon}. \tag{8.20}$$

The mean square end-to-end separation can be found from

$$\langle R_N^2(T, \varepsilon)\rangle = \frac{1}{m}\sum_i w_i R_{N,i}^2\, \mathrm{e}^{-v_i\beta\varepsilon}(Z)^{-1}$$

where w_i is any weight introduced by survival biasing etc.

If ε is negative, corresponding to attraction between the segments in the chain, chains with large values of v_i will get the greatest weightings in the sums and such configurations correspond to chains relatively tightly packed compared to chains in the absence of this interaction ($\varepsilon = 0$). In general, the dependence of $\langle R_N^2 \rangle$ on N will also be different in this case and will depend on $\beta\varepsilon$. It is found that at a certain value of $\beta\varepsilon$, $\beta\varepsilon = z_0$, the dependence is just the same as for chains without any interaction whatsoever, i.e. no excluded volume effect, $\sqrt{\langle R^2 \rangle} \sim N^{1/2}$. The temperature at which this occurs, ε/kz_0, is known as the Θ or Flory temperature. At this temperature the coiling tendency due to the van der Waals attraction between the atomic subunits just compensates for the repulsion induced by the excluded volume effect, so that the average length of a molecule, and its configurational properties in general, are just what would be obtained for a pure Markovian random walk.

Although the most direct manifestation of the temperature is the N-dependent behaviour of $\langle R_N^2 \rangle$, the testing of it would involve the generation of many independent walks at different N and β. In the anticipation that not only this property, but other properties of the chain at this temperature will mimic those of a pure Markovian walk, you should consider walks of just one length, e.g. N in the region 70–100, and for different values of $\beta\varepsilon$ (in the range 0–0.5, where 0 corresponds just to the case of excluded volume) compare some average parameters of the walks with their Markovian values. Examples might be the ratio $\langle R_N^2\rangle/\langle S_N^2\rangle$, which has a value of six, or the parameter t in expression (8.18) which has a value of two.

The mean interaction energy in a chain is $E = \langle v\rangle\varepsilon$, so the associated specific heat, cf (8.10), is $Nk_B\beta^2\sigma_E^2$. Hence, a plot of $C/Nk_B\varepsilon^2 = \beta^2\sigma_v^2$ against $\beta\varepsilon$ will show if there is any behaviour analogous to a phase transition suggested in the neighbourhood of the Θ temperature.

8.7 The Monte Carlo Method in Quantum Mechanics

In recent years widespread use has been made of Monte Carlo methods in the field of high energy physics to evaluate the consequences of a theory for the structure of hadronic matter, known as quantum chromodynamics (QCD). In this theory, hadronic particles, e.g. π mesons or nucleons, are viewed as being composed of quarks which interact via quanta known as gluons. Both the sources (quarks) and the quanta (gluons) are assumed to carry several new charges, called colours, which play a role in determining the interactions between the subunits. To investigate predictions of the model the motion of individual quarks, determined by an appropriate Hamiltonian, may be simulated. In these simulations, instead of a continuum of values of coordinates of a particle, it is found convenient to restrict possible locations to a lattice of points in space–time. The world line of a quark is then approximated by a sequence of steps between nearest-neighbour points on the four-dimensional lattice, with step sizes small compared to hadronic dimensions and interaction times. Under these conditions, such calculations must be carried out in the framework of quantum mechanics.

Within this framework a natural formalism to describe the evolution of a system exists, namely the formalism of path integrals due to Feynman. As we shall see, a practical way of calculating these path integrals is the Monte Carlo method. This is so because there is a very close relationship between the form of these path integrals and the partition function in statistical mechanics; with the quarks confined to discrete lattice points, their possible paths play a role analogous to the configurations of spins in the Ising model discussed in Project 8C above. In the Ising model the role of different configurations of lattice spins was determined by the temperature; how can the temperature enter into determining the values of the fields? Of course, it doesn't; the bridge between the statistical mechanical problem and that for interacting fields depends on a fortunate mathematical accident, namely that the path integrals occurring in Feynman's path integral formulation are identical in form to the partition function in statistical mechanics, with the coupling constant (strength of the potential) replacing $\beta = 1/k_BT$. The path integral formalism refers to four space–time, rather than three space, dimensions and the sites in this space correspond to quark locations, while the analogues of the α_i are field variables. The interpretation of the resulting calculations also has features in common with the case of statistical mechanics. Concepts like that of a phase transition may be extended to understand the behaviour of quark matter at high energy densities, e.g. in the collision between two hadrons.

The field variables defined at a lattice point in such theories, because of the several 'charges' (colours) involved as well as the importance of the occurrence of virtual particles in the very short time steps, must be multicomponent quantities. The ideas behind the method are best illustrated by reference to a much simpler, non-relativistic case, which we can do by considering its application to the two-dimensional space–time evolution of a particle in a one-dimensional potential. We shall see how to obtain the ground state energy and wavefunction, quantities which would ordinarily be found by solving the Schrödinger equation, but first the connection between the two viewpoints must be established.

Schrödinger's equation is a differential equation which determines the evolution of a probability amplitude, $\Psi(x, t)$, for a system; it satisfies the following equation, where H is the Hamiltonian of the system,

$$H\Psi(x, t) = \mathrm{i}\hbar\, \partial\Psi/\partial t. \tag{8.21}$$

With the initial state, $\Psi(x, t_0)$, specified the solution of the equation determines the amplitude, and thus the probability density, $|\Psi(x, t)|^2$, at later times. In the Feynman viewpoint, this probability amplitude at a given time is viewed as arising from its propagation from all locations of the initial state, that is

$$\Psi(x, t) = \int K(x, t; x_0, t_0)\Psi(x_0, t_0)\, \mathrm{d}x_0. \tag{8.22}$$

The function $K(x, t; x_0, t_0)$, known as a propagator, represents the strength with which the value of the initial wavefunction at x_0 influences its value at x at the later time. By expanding the solution of (8.21) in terms of time-independent eigenfunctions, $\psi_n(x)$, of the operator H, we can write, formally,

$$\Psi(x, t) = \sum_n c_n \exp\left(-\frac{\mathrm{i}}{\hbar} E_n t\right)\psi_n(x). \tag{8.23}$$

It can be shown (see, for example, §13 of Matthews 1968) that the coefficients in this expansion are given by

$$\mathrm{c}_n = \int \mathrm{d}x_0 \Psi(x_0, 0)\psi_n^*(x_0). \tag{8.24}$$

When these are inserted into (8.23) and a comparison with (8.22) is made, we find the equivalence

$$K(x, t; x_0, t_0 = 0) = \sum_n \psi_n^*(x_0)\psi_n(x) \exp(-\mathrm{i}E_n t/\hbar).$$

If we now assume that this identity is valid when extrapolated to imaginary values of t, it becomes, writing $t = -\mathrm{i}\tau$,

$$K(x, -\mathrm{i}\tau; x_0, 0) = \sum_n \psi_n^*(x_0)\psi_n(x) \exp(-E_n \tau/\hbar). \tag{8.25}$$

At sufficiently large values of τ, the exponential on the right-hand side of this equation will ensure that the term with the smallest energy, the ground state, will dominate the sum. If we take $x_0 = x$ and ignore the later terms, we can write

$$K(x, -\mathrm{i}\tau; x, 0) \simeq |\psi_0(x)|^2 \exp(-E_0\tau/\hbar)$$

where E_0 is the energy of the ground state. Specifically, when $\tau \gg \hbar/(E_1 - E_0)$, where E_1 is the energy of the first excited state, we have

$$|\psi_0(x)|^2 = \exp(E_0\tau/\hbar)K(x, -\mathrm{i}\tau; x, 0). \tag{8.26}$$

By requiring normalisation, i.e. $\int |\psi_0(x)|^2 \,\mathrm{d}x = 1$, we see that the squared modulus of the ground state wavefunction can be expressed in terms of the propagator,

$$|\psi_0(x)|^2 = \lim_{\tau \to \infty} \left[K(x, -\mathrm{i}\tau; x, 0)\left(\int_{-\infty}^{\infty} K(x, -\mathrm{i}\tau; x, 0) \,\mathrm{d}x \right)^{-1} \right]. \tag{8.27}$$

We need to know how to evaluate the propagator.

The propagator has contributions from all visualisable paths in space–time linking the initial, (x_0, t_0), and final, (x, t), points with the contribution on each such path being given by a simple phase factor $\exp[(\mathrm{i}/\hbar)S_{\text{path}}]$, where S_{path} is the classical action evaluated along the path, that is

$$S_{\text{path}} = \int_{t_0\ \text{path}}^{t} L \,\mathrm{d}t.$$

Here, L is the Lagrangian of the system. If we denote an individual path by $X_j(t)$, we can write

$$K(x, t; x_0, t_0) = \sum_j \exp[(\mathrm{i}/\hbar)S[X_j]] \qquad t > t_0. \tag{8.28}$$

For the case of a free particle this is shown in Chapter 3 of Baym (1969). The standard introduction to the topic is Feynman and Hibbs (1965). To see how this summing over paths may be reduced to a multiply infinite integral we divide the interval $(t - t_0)$ into $(N + 1)$ parts of equal duration ε, i.e. we write

$t_j = t_0 + j\varepsilon$ and $t_{N+1} = t$. Then, using (8.22), we can write

$$\begin{aligned}\Psi(x, t) &= \int K(x, t_{N+1}; x_N, t_N)\Psi(x_N, t_N)\,\mathrm{d}x_N \\ &= \int K(x, t_{N+1}; x_N, t_N) \int K(x_N, t_N; x_{N-1}, t_{N-1}) \\ &\quad \times \Psi(x_{N-1}, t_{N-1})\,\mathrm{d}x_{N-1}\,\mathrm{d}x_N \\ &= \int \mathrm{d}x_1 \ldots . \mathrm{d}x_N \prod_{j=0}^{N} K(x_{j+1}, t_{j+1}; x_j, t_j)\Psi(x_0, t_0)\,\mathrm{d}x_0\end{aligned}$$

where we have written $x = x_{N+1}$. On comparison with (8.22) we see that

$$K(x, t; x_0, t_0) = \int \mathrm{d}x_1 \ldots \mathrm{d}x_N \prod_{j=0}^{N} K(x_{j+1}, t_j + \varepsilon; x_j, t_j).$$

Although the individual propagators in the product themselves require an infinity of paths for their evaluation, it can be shown (Feynman and Hibbs 1965, Mannheim 1983) that for a sufficiently short time interval, ε, only the contribution from one elementary path, the classical path, contributes significantly, and that this contribution is

$$K(x_{j+1}, t_j + \varepsilon; x_j, t_j) = A \exp[(\mathrm{i}/\hbar)S[x_j, x_{j+1}]].$$

Here, $S[x_j, x_{j+1}]$ is the value of the action on the classical path between (x_j, t_j) and $(x_{j+1}, t_j + \varepsilon)$, which can be approximated by a straight line segment, while the value of the constant A, $A = (m/\mathrm{i}h\varepsilon)^{1/2}$, will not be of direct interest to us. Inserting this into the above equation we have the result that

$$K(x, t; x_0, t_0) = A^N \int \mathrm{d}x_1 \ldots . \mathrm{d}x_N \exp[(\mathrm{i}/\hbar)S[x_0, x]]. \qquad (8.29)$$

Here, $S[x_0, x]$ is the total action along a path $X \equiv \{(x_0, t_0), (x_1, t_0 + \varepsilon), \ldots, (x_N, t_0 + N\varepsilon), (x, t)\}$, being the sum of the contributions on the individual segments.

When dealing with non-relativistic motion involving potentials which do not depend on velocity, the Lagrangian can be written as the difference between the kinetic and potential energy. In terms of $\tau = \mathrm{i}t$ it becomes

$$\begin{aligned}L(x, \dot{x}, t) &= -\frac{m}{2}\left(\frac{\mathrm{d}x}{\mathrm{d}\tau}\right)^2 - \Phi(x) \\ &= -E(x, \tau)\end{aligned}$$

i.e. the Lagrangian at imaginary values of t is just the negative of the total energy (or Hamiltonian) evaluated for real values of τ. The action then becomes

$$S[x_j, x_{j+1}] = \int_{t_j}^{t_{j+1}} L(x, \dot{x}, t)\,\mathrm{d}t = \mathrm{i}\int_{\tau_j}^{\tau_{j+1}} E(x, \tau)\,\mathrm{d}\tau.$$

Writing (8.29) in terms of τ, we can rewrite (8.27) as

$$|\psi_0(x)|^2 = \underset{\tau \to \infty}{\text{limit}} \int dx_1 \ldots . dx_N \left[\exp\left(-\frac{1}{\hbar} \int_0^\tau E \, d\tau \right) \right] Z^{-1} \tag{8.30}$$

where

$$Z = \int dx \int dx_1 \ldots . dx_N \exp\left(-\frac{1}{\hbar} \int_0^\tau E \, d\tau \right).$$

The integral in the exponential is to be evaluated along the path ($x_0 = x$, $x_1(\varepsilon)$, $x_2(2\varepsilon)$, ..., $x_N(N\varepsilon)$, $x_{N+1} = x$). The replacement of a single path, figure 8.2(*a*), by a path segmented into equal time intervals is shown in figure 8.2(*b*). Also shown there are two of the remaining infinity of paths which are allowed for in the integration over dx_j. Along any such path we can approximate $(1/\hbar) \int_o^\tau E \, d\tau$ by the expression

$$\begin{aligned} \frac{1}{\hbar} \int_0^\tau E \, d\tau &= \frac{\varepsilon}{\hbar} \sum_{j=0}^{N} \left[\frac{m}{2} \left(\frac{x_{j+1} - x_j}{\varepsilon} \right)^2 + \Phi(x_j) \right] \\ &= \frac{\varepsilon}{\hbar} E(x, x_1, \ldots, x_N) \end{aligned} \tag{8.31}$$

where x_0 and x_{N+1} are set equal to x. In this way, an energy is associated

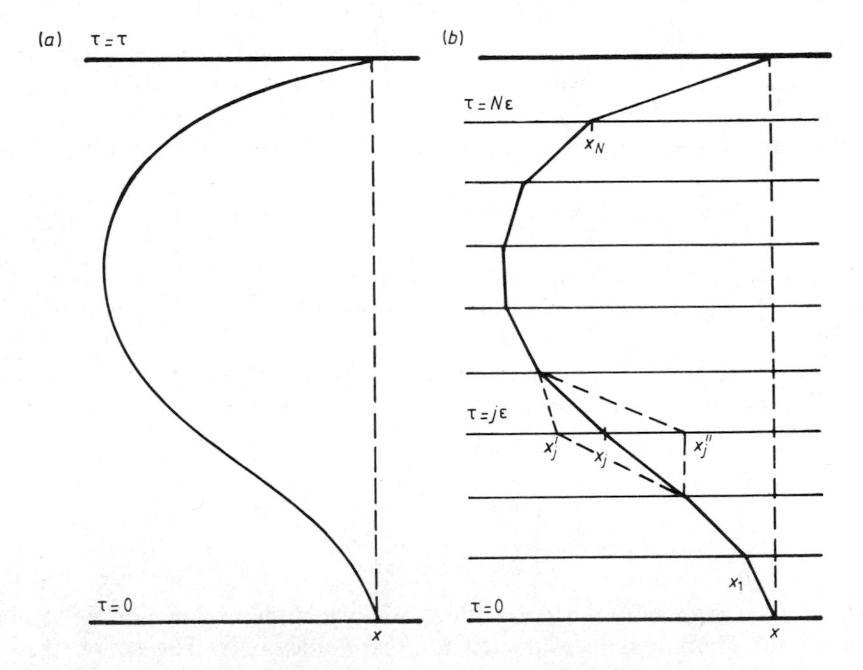

Figure 8.2 (*a*) One of the infinity of possible paths linking $(x, 0)$ to (x, τ). (*b*) A segmented approximation to this path and, in addition, two paths (via x_j' and x_j'') which would be included in the integration at time τ_j.

with every segmented path. Expression (8.30) then takes the form

$$|\psi_0(x)|^2 = Z^{-1} \int \mathrm{d}x_1 \ldots \mathrm{d}x_N \exp\left(-\frac{\varepsilon}{\hbar} E(x, x_1, \ldots, x_N)\right).$$

If we write $x = x_0$ and integrate over x_0 with a delta function $\delta(x_0 - x)$ inserted in the integrand, we have the equivalent expression,

$$|\psi_0(x)|^2 = \int \mathrm{d}x_0\, \mathrm{d}x_1 \ldots \mathrm{d}x_N\, \delta(x_0 - x) Z^{-1} \exp\left(-\frac{\varepsilon}{\hbar} E(x_0, x_1, \ldots, x_N)\right) \tag{8.32}$$

with the 'partition function' Z given by

$$Z = \int \mathrm{d}x_0\, \mathrm{d}x_1 \ldots \mathrm{d}x_N \exp\left(-\frac{\varepsilon}{\hbar} E(x_0, x_1, \ldots, x_N)\right). \tag{8.33}$$

Expressions (8.32) and (8.33) immediately call to mind equation (8.9), i.e. $|\psi_0(x)|^2$ may be thought of as the value of a function, $\delta(x - x_0)$, averaged over many configurations of the variables x_j (each configuration defining a path), whose probability distribution is just the Boltzmann distribution, (8.7), with $k_B T$ replaced by $\hbar/\varepsilon$. The Metropolis algorithm for selecting configurations according to this distribution may thus be employed, as described in Project 8C above, to evaluate the integral (8.32). How the wavefunction may be evaluated at many different values of x without much extra effort, as well as the details particular to a specified potential, are discussed in Project 8F below.

As a practical method for solving for the ground state wavefunction of a system in one dimension, one can see little merit in the approach outlined above. However, in more advanced quantum mechanical calculations the approach of path integrals has decided advantages. These arise principally from the fact that it is an attempt at an exact calculation, i.e. the potential is inserted into the action integral and a solution sought, rather than as in conventional field theory calculations where its effects are treated as perturbations on a vacuum state. This latter approach works well with electrodynamics where the coupling constant is small but is of little value for interactions such as the strong interaction, where the matrix elements cannot be expanded as a series in the coupling constant which converges. Another advantage arises from putting space–time on to a lattice in a discrete form. Already we have seen that this is helpful when using the Markov chain method to select representative states; it enables a concise specification of the fields to be given in terms of arrays, in addition, it enables a cut-off to be placed on the integrals, corresponding to renormalisation in perturbation theory, before extrapolating from the lattice to the continuum.

*PROJECT 8F: THE GROUND STATE WAVEFUNCTION FOR A SIMPLE HARMONIC OSCILLATOR

In this project, equation (8.32) is to be evaluated using the Metropolis algorithm for a particle of mass m in a one-dimensional, simple harmonic potential, i.e. $\Phi(x) = m\omega^2 x^2/2$ (see Lawande *et al* (1969) and MacKeown (1985)). Taking $\sqrt{\hbar/m\omega}$ for the unit of length and $1/\omega$ for the unit of τ, expression (8.31) takes the form

$$\frac{\varepsilon}{2}\sum_{j=0}^{N}\left[\left(\frac{x_{j+1}-x_j}{\varepsilon}\right)^2 + x_j^2\right] \equiv \varepsilon E(x_0, \ldots, x_N). \qquad (8.34)$$

For an arbitrary initial path linking the $(N+2)$ time levels, with $x_{N+1} = x_0$, this energy can be evaluated, and thereafter successive paths chosen, each differing from the previous one by, at most, the segments common to one time level, e.g. τ_k, using the method described in §7.4 and Project 8C above. A random coordinate x_k, is changed to the value x'_k with a probability $W_{kk'} = \min[1, \exp(-\varepsilon\Delta E)]$, where ΔE is the difference between the energies of the paths incorporating x'_k and x_k, as determined by (8.34). When the 'new' path is chosen (it may be identical to the previous one), the function being integrated, in this case $\delta(x - x_0)$, is to be evaluated and added into a running sum; this sum will be eventually divided by the total number of paths sampled to find its average. If we consider a typical step in such a walk, with the move x_k to x'_k successful (as shown by dotted lines in figure 8.3), the new path, like the old one, will make no contribution to the integral if, as in the figure, x_0 does not equal x. However, this new path, passing through C, can be used in a calculation of $|\psi_0(x'_k)|^2$ because C may be thought of as the starting point of a path CBD whose extension BD, shown as a broken line in figure 8.3, is just a replication of the earlier segments AD. This is so because the value of (8.34) is the same for the two paths. In this way, the wavefunction for all points can be established using the same walk. If the space points are discrete, as in figure 8.3, a sum is to be associated with each such point which is augmented by unity each time that point is selected to determine another path (which may be identical to the previous path). When sufficient steps of the walk have been carried out, the sum corresponding to the point x_i divided by the total number of steps executed is $|\psi_0(x_i)|^2$.

The value of τ which is sufficient to ensure the validity of (8.27) can only be found by trial and error; with the units employed here a value in the range 10–16 should be good enough (MacKeown 1985). The range of x values for which the wavefunction will be appreciable must also be determined empirically, the range -3 to $+3$ in the units used here is suggested. A mesh of $N \times M$ points with spacing ε and h must then be imposed on the $\tau - x$ plane and an initial path linking $x(0) = x(N+1) = 0$ selected. As in the case of the configurations of spins in a similar study of the Ising model in Project 8C, the final result should be independent of the initial configuration chosen, a

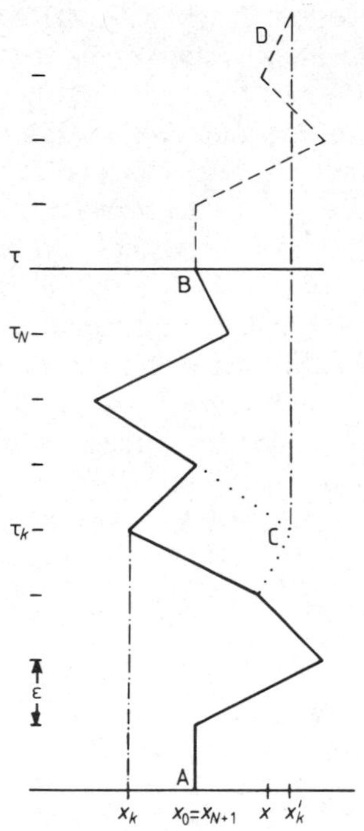

Figure 8.3 The solid line shows one of the path configurations occurring in the evaluation of the integral (8.32). The choice of another path via C (shown dotted) makes a non-zero contribution to that integral only for a value of $x = x'_k$ since x'_k may be thought of as the starting and ending point of a path CBD on which the action is just the same as on ACB.

possible example is the one with all $x_i(\tau) = 0$. To eliminate the influence of this arbitrary initial path a (large) number of steps at the beginning of the walk should be executed, before starting to evaluate the integral by the method discussed above. Successive graphical monitoring of the path after a fixed number of steps may be helpful in deciding when equilibrium configurations are being sampled (continuous monitoring will probably slow down the program to an unacceptable level).

When the wavefunction has been determined, the energy of the ground state may be obtained as the expectation of the Hamiltonian, that is

$$\frac{E}{\hbar\omega} = \frac{1}{2}\int \psi_0^* \left(-\frac{\partial^2}{\partial x^2} + x^2 \right) \psi_0 \, \mathrm{d}x$$

where, because there are no nodes in a ground state wavefunction, we have $\psi_0(x) = (|\psi_0(x)|^2)^{1/2}$. Using the finite difference replacement (3.3) for the operator, the above expression can be evaluated using the values of the wavefunction obtained at the discrete points x_i.

Once developed, a program which will perform these calculations can easily be modified to deal with other potentials, e.g. the square-well potential or Morse potential (see Lawande *et al* (1969)).

For the particular case of the harmonic oscillator potential, as well as for other potentials which can be expressed as quadratics in the coordinates, the use of the Metropolis algorithm, as described above, for selecting the paths is not essential. As we shall now show, they can be selected on the basis of the Gaussian distribution, in which case the technique of over-relaxation, described in Chapter 3 in connection with the solution of finite difference equations, may be employed to accelerate the rate at which paths relax to configurations typical of equilibrium (Whitmer 1984). The previously developed program should be modified along these lines and the speed of the two versions compared.

If a single coordinate, x_k, where $k \neq 0$, is isolated in the expression (8.34) the argument in the exponential assumes the form

$$ax_k^2 + bx_k + c$$

where $a = (2 + \varepsilon^2)/2\varepsilon$, $b = -(x_{k-1} + x_{k+1})/\varepsilon$ and the term c includes all coordinates other than x_k. When exponentiated the probability distribution may be written in the form

$$A \exp[-(x_k - \mu_k)^2/2\sigma^2] \tag{8.35}$$

with $\mu_k = (x_{k-1} + x_{k+1})/(2 + \varepsilon^2)$, $\sigma^2 = \varepsilon/(2 + \varepsilon^2)$ and A is a function only of coordinates other than x_k. Working with a continuum of variables $x(k\varepsilon)$, rather than discrete values as described earlier, a new value $x'(k\varepsilon)$ is chosen in a given step in the Markov chain by selecting from a Gaussian distribution with mean μ_k and variance σ^2, using one of the methods described in §7.4. We note that the mean, μ_k, to which the coordinate $x(k\varepsilon)$ tends to relax, itself depends on the adjacent coordinates on the path, $x[(k-1)\varepsilon]$ and $x[(k+1)\varepsilon]$, and so changes with the evolution of the walk.

The role of the coordinates of neighbouring points on the path in determining the most recent value of the coordinate at a given time step is reminiscent of the role played by the value of a function at neighbouring points in determining the new value to be assigned when using relaxation to solve the difference equations arising from an elliptical differential equation, described in §3.3. However, we also saw in that section that if the influence of these latest neighbouring values was suppressed somewhat, and some memory of the previous value of the function at the point retained, a more rapid convergence to a solution was obtained. This is the method of successive over-relaxation (SOR) described in §3.5. This same method can be employed in selecting paths when stepping through the Markov chain; instead of the value μ_k given above, the mean of the Gaussian distribution from which the coordinate $x'(k\varepsilon)$ is chosen is taken to be $\mu'_k = \omega\mu_k + (1 - \omega)x(k\varepsilon)$, where ω is a relaxation parameter $(0 < \omega < 2)$ whose optimum value has to be determined. In this case (Adler 1981), the variance of the distribution should also be changed to $(\sigma')^2$, where $(\sigma')^2 = \omega(2-\omega)\sigma^2$, i.e. (8.35) is replaced by

$$A \exp[-(x'_k - \mu'_k)^2/2(\sigma')^2].$$

This has the effect, on average, of relaxing a coordinate x_k to a value μ'_k. This relaxation, in minimising the contribution to the energy, or action, of the path in its neighbourhood, also makes allowance for the fact that the neighbouring coordinates x_{k-1} and x_{k+1} will not, in general, lie on a path which minimises the energy over the whole path.

An investigation of the optimum value of ω, in the manner described in Adler (1981), should be carried out.

Exercises

8.1 A better approximation to the neutron energy distribution in fission is the two-parameter *Watt spectrum*,

$$p(E)\,\mathrm{d}E = A\,\mathrm{e}^{-E/T} \sinh\sqrt{bE}\,\mathrm{d}E$$

where $0 \leqslant E < \infty$, $T = 0.965$ MeV, $b = 2.29\ \mathrm{MeV}^{-1}$ and A is a normalising constant. Verify that the following application of an extension of the rejection technique discussed in §7.4 (due to Kalos) selects variates from this distribution. Select two variates from the uniform distribution, ξ_1 and ξ_2. Defining $K = 1 + (bT/8)$, $L = T(K + \sqrt{K^2 - 1})$ and $M = K - 1 + \sqrt{K^2 - 1}$, set $E = -L \ln \xi_1$ if $-bL \ln \xi_1 \geqslant [M(1 - \ln \xi_1) + \ln \xi_2]^2$.

8.2 Verify the following procedure as a method for selecting the components $(\omega_1, \omega_2, \ldots, \omega_N)$ of an isotropically distributed unit vector in an N-dimensional space. Let $k = N/2$ or $(N+1)/2$ according to whether N is even or odd, select k pairs of variates from the uniform distribution

$$(\xi_{2i-1}, \xi_{2i}) \qquad i = 1, \ldots, k$$

and use them to form

$$v_{2i-1} = (-\ln \xi_{2i-1})^{1/2} \cos 2\pi\xi_{2i}$$

$$v_{2i} = (-\ln \xi_{2i-1})^{1/2} \sin 2\pi\xi_{2i}.$$

Then for $j = 1$ to N, set $\omega_j = v_j/\rho$ where

$$\rho = \left(\sum_{j=1}^{N} v_j^2\right)^{1/2}.$$

(See also Marsaglia (1972).)

8.3 The eight configurations of a system of three particles, each having two possible states, is shown below, along with the interaction energies of each, Φ_i, expressed in units of kT. Assuming the occurrence of each state is governed by the Boltzmann distribution, i.e. $p_i = e^{-\Phi_i}/Z$ where $Z = \Sigma\, e^{-\Phi_i}$, show that the transition matrix, **W**, determined by the Metropolis algorithm (expression (7.14)) is as indicated. Write a program which, starting with an arbitrary state, will use this matrix to generate successive states and compare the frequency of occurrence of the different states with the Boltzmann distribution. The influence of the initial state should be investigated.

i	State	Φ_i	**W**							
1	↑↑↑	3	0.25	0.25	0.25	0.25	0	0	0	0
2	↑↑↓	2	0.092	0.408	0	0	0	0.25	0.25	0
3	↑↓↑	2	0.092	0	0.408	0	0.25	0.25	0	0
4	↓↑↑	2	0.092	0	0	0.408	0.25	0	0.25	0
5	↓↓↑	1	0	0	0.092	0.092	0.566	0	0	0.25
6	↑↓↓	1	0	0.092	0.092	0	0	0.566	0	0.25
7	↓↑↓	1	0	0.092	0	0.092	0	0	0.566	0.25
8	↓↓↓	0	0	0	0	0	0.092	0.092	0.092	0.724

8.4 Show that the square of the radius of gyration may be written in the alternative form of equation (8.16). If the chain is confined to a simple cubic lattice of unit side, show that an individual term R_i^2 in the summation in that equation may be written as

$$R_i^2 = R_1^2 + 2\sum_{j=1}^{i-1} (x_j \Delta x_j + y_j \Delta y_j + z_j \Delta z_j) + (i-1)$$

where Δx_j, Δy_j, Δz_j are the components of the jth step of the walk. In practice, two of these coordinates are zero while the third has the value ± 1. Accumulating the quantities in this way will be more efficient than summing the squares (see McCrackin *et al* (1973)).

8.5 Show that for an unconstrained random walk in three dimensions the relationship between the square of the radius of gyration and the mean square length given in §8.5 holds for large N, i.e. $\langle R^2 \rangle / \langle S^2 \rangle = 6$. (See §14-2 of McQuarrie (1973).)

8.6 Develop a program to estimate the quantity v in expression (8.19) for a simple cubic lattice.

References

Adler S L 1981 *Phys. Rev.* D **23** 2901–4
Baym G 1969 *Lectures on Quantum Mechanics* (New York: W A Benjamin)

Chandrasekhar S 1943 *Rev. Mod. Phys.* **15** 2–89

Davis D H 1963 in *Methods in Computational Physics* ed. B Adler *et al* vol. 1 p 67–88 (New York: Academic)

Ehrman J R, Fosdick L D and Handscomb D C 1968 *J. Math. Phys.* **1** 547–58

Feynman R P and Hibbs A R 1965 *Quantum Mechanics and Path Integrals* (New York: McGraw-Hill).

Fox L 1962 *Numerical Solutions of Ordinary and Partial Differential Equations* (Oxford: Pergamon)

Hammersley J M and Handscomb D C 1964 *Monte Carlo Methods* (London: Chapman and Hall)

Henry A F 1975 *Nuclear Reactor Analysis* (Cambridge, MA: MIT Press)

Landau D P 1976 *Phys. Rev.* B **13** 2997–3011

Lawande S V, Jensen C A and Sahlin H L 1969 *J. Comput. Phys.* **3** 416–43

MacKeown P K 1985 *Am. J. Phys.* **53** 880–5

Mandl F 1971 *Statistical Physics* (London: Wiley)

Mannheim P D 1983 *Am. J. Phys.* **51** 328–34

Marsaglia G 1972 *Ann. Math. Stat.* **43** 645–6

Matthews P T 1968 *Introduction to Quantum Mechanics* 2nd edn (London: McGraw-Hill)

Mazur J and McCrackin F L 1968 *J. Chem. Phys.* **49** 648–65

McCrackin F L, Mazur J and Guttman C M 1973 *Macromolecules* **6** 859–71

McQuarrie D A 1973 *Statistical Thermodynamics* (New York: Harper and Row)

Mendelson M R 1968 *Nucl. Sci. Eng.* **32** 319–31

Sauer G 1981 *Am. J. Phys.* **49** 13–19

Wall F T and Erpenbeck J J 1959 *J. Chem. Phys.* **30** 634–7

Whitmer C 1984 *Phys. Rev.* D **29** 306–11

Wood W W 1968 in *Physics of Simple Liquids* ed. H N V Temperley, J S Rowlinson and G S Rushbrooke (Amsterdam: North-Holland) Chapter 5

Appendix

Mathematical Background

A.1 Statistical Concepts

Many physical quantities, unlike, for example, the charge on the electron, cannot be characterised by a unique value which is obtained in every measurement. Examples are the number of hours of sunshine in a day, the distance a 10 MeV neutron will travel in water before it interacts, or the reading that will be obtained when dice are tossed. Such quantities are called *random* variables. The possible value for such a quantity, denoted by x, although not unique, is not unconstrained, rather it is distributed in a way which can usually be approximated by a simple mathematical function called the *probability density function* (or frequency function), $p(x)$. This means that the probability of obtaining a value of x in the interval $(x, x + \mathrm{d}x)$—or a value x_i if it is a discrete variable—is given by $p(x)\,\mathrm{d}x$ or $p(x_i)$ respectively, with the condition

$$\int_{-\infty}^{\infty} \mathrm{d}x\, p(x) = 1 \qquad \left(\text{or} \sum_i p(x_i) = 1\right).$$

The corresponding cumulative distributions, i.e. the probability of obtaining a value less than a specified x, are

$$P(x) = \int_{-\infty}^{x} \mathrm{d}x\, p(x) \qquad \left(P(x_j) = \sum_{i=1}^{j} p(x_i)\right). \tag{A.1}$$

Important parameters specifying a distribution are its *mean*, μ, and *variance*, σ^2, defined by

$$\mu = \int \mathrm{d}x\, x\, p(x) \tag{A.2}$$

$$\sigma^2 = \int \mathrm{d}x (x - \mu)^2 p(x) \equiv \langle x^2 \rangle - \mu^2. \tag{A.3}$$

Here, $\langle x^2 \rangle$ is the second moment,

$$\langle x^2 \rangle = \int \mathrm{d}x\, x^2 p(x).$$

The quantity σ is known as the *standard deviation* of the distribution.

If an event, or physical quantity, is characterised by two values, say x and y, both of which are distributed, an overall distribution function can be defined in an analogous way having means μ_x and μ_y and variances σ_x^2 and σ_y^2. In addition, it is useful in this case to define a quantity, called the covariance of x and y, by

$$\mathrm{cov}(x, y) = \int \mathrm{d}x\, \mathrm{d}y(x - \mu_x)(y - \mu_y)p(x, y). \tag{A.4}$$

If the variables are *independent*, i.e. $p(x, y)$ is such that $\int \mathrm{d}y(y - \mu_y)p(x, y)$ vanishes, the covariance is zero, although the converse is not in general true—vanishing covariance does not imply independence. The *correlation coefficient*, $\rho \equiv \mathrm{cov}(x, y)(\sigma_x^2 \sigma_y^2)^{-1/2}$, lies in the region $-1 \leqslant \rho \leqslant 1$.

It is often the case that $p(x)$ is not known a priori and has to be inferred from the observed values, e.g. from the data on the hours of sunshine for many days. In mentally constructing a distribution function for the quantity x we have assumed unique values for μ and σ^2, characterising the whole population of such actual and possible events, but we do not, in general, know their values. The best we can do is to estimate them from the finite sample of readings obtained, say N. Thus we distinguish between the mean and variance of the underlying distribution—the population—and our estimate of them based on the finite sample of size N. Optimum estimates are given by

$$m = \sum_1^N x_i/N \tag{A.5}$$

for the mean and

$$s^2 = \sum_1^N (x_i - m)^2/(N - 1) \tag{A.6}$$

for the variance.

How meaningful these estimates are depends on the distribution $p(x)$, and on the assumption that the values were drawn at random. Although this is obviously an important question, we shall not pursue it here.

A.2 Useful Distributions

Here, we summarise the properties of some distributions which are useful in physics, distinguishing between the cases of continuous and discrete variables.

The **Gaussian (normal) distribution** takes the form

$$p(x)\, \mathrm{d}x = \frac{1}{\sigma\sqrt{2\pi}} \exp\left[-\frac{(x - \mu)^2}{2\sigma^2}\right] \mathrm{d}x \tag{A.7}$$

where the mean, μ, and variance, σ^2, are explicitly shown. When dealing with this distribution it is often convenient to work in terms of a standardised variable, t, such that individual observations can be written as $x_i = \mu + \sigma t_i$. This variable is distributed as

$$p(t)\,dt = (2\pi)^{-1/2}\,e^{(-t^2/2)}\,dt \tag{A.8}$$

where $t = (x - \mu)/\sigma$ and has a mean equal to zero and a variance of one. Tables of the cumulative distribution $P(t)$ are widely available, e.g. Abramowitz and Stegun (1965). From them, we can see that a variable satisfying the normal distribution has a probability of 50%, 68.3%, 95.4% and 99.7% of occurring within $\pm 0.68\sigma$, $\pm\sigma$, $\pm 2\sigma$ and $\pm 3\sigma$ from the mean, respectively.

The **exponential distribution**,

$$p(x)\,dx = (1/\mu)\,e^{-x/\mu}\,dx \tag{A.9}$$

with mean μ and variance μ^2, is familiar from radioactive decay.

Useful functions for parametrising data are the following two distributions. The **beta distribution** is

$$p_\beta(x)\,dx = \frac{\Gamma(a+b)}{\Gamma(a)\Gamma(b)}\,x^{a-1}(1-x)^{b-1}\,dx \qquad 0 \leqslant x \leqslant 1 \tag{A.10}$$

where a and b are greater than 1, $\mu = a/(a+b)$ and

$$\sigma^2 = \frac{ab}{(a+b+1)(a+b)^2}.$$

$\Gamma(x)$ is the gamma function.

The other useful function is the **gamma distribution**, given by

$$p_\gamma(x)\,dx = (\Gamma(a)b^a)^{-1}\,e^{-x/b}\,x^{a-1}\,dx \qquad x \geqslant 0 \tag{A.11}$$

with $\mu = ab$ and $\sigma^2 = ab^2$. An example of the latter is the Maxwell–Boltzmann velocity distribution.

Another useful distribution is the **uniform distribution**,

$$p(x)\,dx = dx/(b-a) \qquad a \leqslant x \leqslant b \tag{A.12}$$

with mean $\mu = (b-a)/2$ and variance $\sigma^2 = (b-a)^2/12$.

Important examples of discrete distributions are the **binomial** and **Poisson distributions**. If the totality of outcomes of an experiment is such that all events can be assigned to one of two mutually exclusive classes, and the probability of the outcome belonging to one of these classes is α, then the probability that in a series of N experiments or trials a total of n events will be obtained in this class is

$$p_N(n) = \frac{N!}{n!(N-n)!}\,\alpha^n(1-\alpha)^{N-n} \qquad 0 \leqslant n \leqslant N. \tag{A.13}$$

In this case, $\mu = N\alpha$, $\sigma^2 = N\alpha(1-\alpha)$ and the distribution is the **binomial distribution**. Note that, here, N is a fixed parameter, not a number of independent samplings from the distribution, as it might be in a histogram based on N observations sampled from the Gaussian distribution. In general, we will have only one sampling, resulting in n out of the N events falling in the particular class. Thus, as an estimate of μ, we take n and therefore, as an estimate of α, we have n/N. The estimate of the variance is then

$$s^2 = n[1 - (n/N)]. \tag{A.14}$$

If the probability of an individual event is very small and the sampling population is very large, i.e. $N \to \infty$ but $\alpha = n/N$, a finite value, the binomial distribution simplifies to the **Poisson distribution**

$$p(n) = \mu^n \mathrm{e}^{-\mu}/n! \tag{A.15}$$

with mean μ and variance $\sigma^2 = \mu$.

For $\mu \gg 1$, the distribution of the variable $(n-\mu)/(\mu)^{1/2}$, which is equivalent to t, tends to the standardised normal distribution, (A.8) above.

It can easily be shown that a variable which is a linear combination of any number, N, of independent random variables, r_i, each of which separately follows a Gaussian distribution, i.e. $z = \Sigma_{i=1}^{N} r_i$, is itself Gaussian-distributed. Subject to certain mild conditions, this is true, for large enough N, for a sum of variables which follow another distribution, or even where the variables in the sum are from different distributions. This is known as the *central limit theorem*. Effectively, $N \gtrsim 10$ will satisfy this in practice, though of course none of the variables must be pathological, e.g. from the Lorentz distribution for which no mean or variance can be defined. This theorem is useful in the interpretation of average estimates arrived at by a Monte Carlo calculation based on a series of random walks, discussed in Chapter 7.

When a variate x is known to be from the Gaussian distribution, or is the sum of several variates such that the central limit theorem applies, the probability of its expectation differing from it by a specified amount ε, given by $P(|x-\mu| > \varepsilon)$, may be read from tables of the cumulative Gaussian function. When the distribution which the variate obeys is not known, we can still say something about this probability. By definition, (A.3), the variance on a quantity, is an integral in which all contributions are positive. If we omit a range $\pm\varepsilon$ of the variable around μ, we therefore have

$$\sigma^2 > \int_{-\infty}^{\mu-\varepsilon} \mathrm{d}x(x-\mu)^2 p(x) + \int_{\mu+\varepsilon}^{\infty} \mathrm{d}x(x-\mu)^2 p(x).$$

Everywhere in these integrals $|x-\mu|^2 \geqslant \varepsilon^2$, so the inequality is strengthened if we write

$$\sigma^2 > \varepsilon^2 \left[\int_{-\infty}^{\mu-\varepsilon} \mathrm{d}x\, p(x) + \int_{\mu+\varepsilon}^{\infty} \mathrm{d}x\, p(x) \right] \equiv \varepsilon^2 P(|x-\mu| > \varepsilon).$$

In other words,

$$P(|x-\mu|>\varepsilon)<\sigma^2/\varepsilon^2 \tag{A.16}$$

i.e. the probability that the observed value differs in magnitude from its expectation by as much as ε is less than σ^2/ε^2. This is known as *Chebyshev's inequality*.

For the case where x is itself the mean, m, of N values of the same variate, independently selected from a population with variance σ^2, the variance on m (see (A.18)) is σ^2/N, and

$$P(|m-\mu|>\varepsilon)<\sigma^2/N\varepsilon^2.$$

The probability may be made arbitrarily small by selecting a 'sufficiently large' N, in which case the sample mean, m, is guaranteed to converge to the population mean, μ. This is known as the *law of large numbers* and is fundamental to all estimation based on sampling.

The sum of squares of a normal deviate, i.e. $z'=\Sigma_{i=1}^{N} r_i^2$, is not normally distributed. Its distribution is known as the **chi-squared distribution** with N degrees of freedom,

$$p_{\chi^2,N}(z')=(z'/2)^{(N/2)-1}\,[2\Gamma(N/2)]^{-1}\exp(-z'/2). \tag{A.17}$$

The estimate, m, of the mean of a set of data is itself a random variable in the sense that if other samples of size N were drawn from the population, different values would be obtained. That is to say, it would follow some distribution, the mean of which, if the m values are unbiased, is the mean of the population. Remembering that

$$m=\sum_{i=1}^{N} x_i/N$$

the central limit theorem tells us that, for 'large N', m is approximately distributed as a Gaussian (true at all N if the x_i values are from a normal population). Also, if all the x_i are independently selected from the *same* population (in which case their covariance vanishes), the variance of the m-distribution will be

$$\sigma_m^2=N\frac{\sigma_{x_i}^2}{N^2}=\frac{\sigma^2}{N}. \tag{A.18}$$

Hence, on the basis of one sample of size N, if we knew σ^2 we could put some confidence limits on the value of the population mean μ, e.g. from tables of the normal distribution, with 90% confidence we can say that μ lies in the range $m-1.64\sigma_m$ to $m+1.64\sigma_m$, or, alternatively, there is only a 0.27% probability that μ lies outside the range $m\pm 3\sigma_m$. If we use our estimate s^2 for σ^2, see expression (A.6), these conclusions will not be rigorous but will often be good enough for our purposes. The rigorous procedure in this case, using the t-distribution, is discussed in Eadie *et al* (1971) and Brandt (1976).

The discussion up until now, and in particular the validity of (A.18) as an estimate of the uncertainty associated with a sample mean, has assumed that the sequence of values x_i for which the mean is calculated has been selected at random from the underlying population. In such a case there would be no significant correlations between values in the sequence. In Monte Carlo calculations, especially those using the Metropolis algorithm for the Markov chain method of selecting variates from a distribution (discussed in Chapters 7 and 8), significant correlations may exist between successive members of the sequence of values calculated. The estimate (A.18) of the variance of the mean of a sample of size N in such a case may seriously underestimate the real uncertainty of this quantity. To allow for correlations, the following method (suggested in Daniell *et al* (1984)) may be used to calculate σ_m. Use (A.5) and (A.6) to calculate m and s^2. For values $k \geqslant 1$ calculate the k-separation correlations, defined by

$$r_k = \frac{2}{N^2} \sum_{i=1}^{N-k} (x_i - m)(x_{i+k} - m)$$

until a value, $k = k_0$, is found which satisfies

$$r_k < 2(N-1)\frac{s^2}{N^2}\frac{(2k+1)^{1/2}}{N^{1/2}}.$$

Write $K = k_0 - 1$. If the condition $K \ll N$ is not satisfied, the sample size N, for the degree of correlation existing in it, is too small to enable any meaningful estimate of the uncertainty of the sample mean to be made. If $K \ll N$ is satisfied, define an overall correlation coefficient R_K as follows:

$$R_K = 0 \qquad \text{if } K = 0$$

$$R_K = \sum_{k=1}^{K} r_k \qquad \text{if } K > 0.$$

The standard deviation on the mean may then be written as

$$\sigma_m = \left[\frac{s^2 + NR_K}{N - 2K - 1}\right]^{1/2}. \qquad \text{(A.19)}$$

We note that if all correlations are negligible, i.e. $K = 0$, this just reduces to (A.18).

A.3 The Chi-squared Test

It is easy to write down many probability density functions as we did above, but, mindful of the fact that we always have a finite sample (often very finite!), how do we know that the physical quantity satisfies any of them? The best we can do is to carry out some test on the sample whose significance we understand. The most straightforward, if not always the most useful, of these tests is the chi-squared test.

In order to test whether a set of observed data, x_i, where $i = 1, \ldots, N$, can be assumed to have been selected from a population distributed as $p(x)$, the x_i values are histogrammed into k bins. Each bin will then contain n_j, where $j = 1, \ldots, k$, items and these are used to form the quantity (often denoted as χ^2)

$$T = \sum_{j=1}^{k} \frac{(n_j - N\alpha_j)^2}{N\alpha_j}. \tag{A.20}$$

Here, α_j is the probability, assuming $p(x)$, of an event occurring in the jth bin, that is

$$\alpha_j = \int_{x_j^-}^{x_j^+} \mathrm{d}x\, p(x)$$

$x_j^{\pm}$ being the limits on this bin as shown in figure A.1. In general, the distribution $p(x)$ will involve some parameters, r in number; e.g. μ and σ in the case of the Gaussian. How this quantity T will be distributed, e.g. over many repetitions of the experiment, depends on whether these r parameters are specified independently of the data n_j or as is often the case, are estimated from the observations n_j themselves. If the data x_i are a sample of a population distributed according to $p(x)$, then under certain circumstances T will be distributed as the chi-squared distribution, equation (A.17), with v degrees of freedom, that is

$$p(T) = p_{\chi^2,v}(T).$$

If no parameters are estimated from the data, $v = k - 1$, while if h $(h \leqslant r)$ parameters are estimated from the (grouped) data in this way, $v = k - 1 - h$.

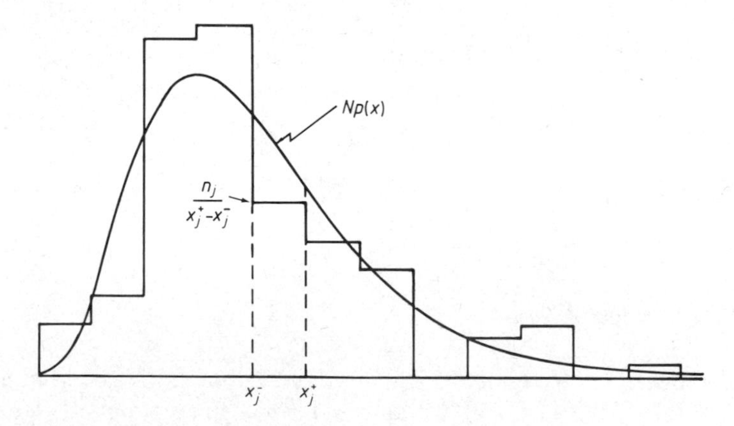

Figure A.1 Graphical comparison of observed numbers, n_j, with predictions based on a distribution $p(x)$.

The conditions referred to above are that the numbers in any one bin should be approximately normally distributed about the expected number in the bin. The actual distribution of these quantities is the binomial distribution and it can only be usefully approximated by the Gaussian if $N\alpha_j$, the expected number, is greater than or equal to five. Binning cells should be chosen so that this condition is satisfied. However, the number of cells must not be too small, there must be enough of them so that the histogram can mimic the variation in the distribution $p(x)$. The optimum number of cells, subject to $N\alpha_j$ being greater than five, is approximately $k = 3N^{2/5}$, and the boundaries of the bins should be chosen such that they contain equal expected numbers, based on $p(x)$. Of course, bins should not be smaller than the degree of resolution in the observed data.

When a value for T is found in this way, the likelihood of finding a value greater than or equal to this value in a random sampling of N variates from the distribution $p(x)$ can be found from tables of the cumulative chi-squared distribution, $P_{\chi^2, \nu}$. It is given by $1 - P_{\chi^2, \nu}(T)$. Values of this likelihood, expressed as a percentage, for several values of ν are shown in table A.1. To use this table the value of the standardised variable u, where $u \equiv (T - \nu)/\sqrt{2\nu}$, should be found; u is a convenient variable both from the point

Table A.1. Percentage probability of obtaining a value larger than u in a random sample, where $u = (T - \nu)/\sqrt{2\nu}$.

u	ν					
	1	2	4	10	20	50
−3.2						99.99
−2.8					99.99	99.98
−2.4					99.98	99.80
−2.0				99.98	99.54	98.88
−1.6				98.49	97.03	95.94
−1.2			96.24	91.43	90.12	89.33
−0.8		81.87	78.39	77.86	77.98	78.22
−0.4	50.99	54.88	58.01	60.82	62.23	63.46
0	31.73	36.79	40.60	44.05	45.79	47.33
0.4	21.08	24.66	27.41	29.94	31.25	32.42
0.8	14.43	16.53	18.04	19.31	19.92	20.42
1.2	10.05	11.08	11.65	11.93	11.95	11.88
1.6	7.09	7.43	7.41	7.10	6.79	6.42
2.0	5.04	4.98	4.66	4.10	3.69	3.24
2.4	3.61	3.34	2.91	2.30	1.92	1.53
2.8	2.59	2.24	1.80	1.27	0.96	0.69
3.2	1.87	1.50	1.10	0.68	0.47	0.29
3.6	1.36	1.01	0.67	0.36	0.22	0.12
4.0	0.99	0.67	0.41	0.19	0.10	0.05

of view of tabulating the distribution, and because the chi-squared distribution in terms of this variable tends to the standardised normal distribution, (A.8), when the number of degrees of freedom is very great. If this probability is small, say less than or equivalent to 5%, there is reasonable suspicion that the population from which the sample of N came is not characterised by the distribution $p(x)$. If, on the other hand, T is very small, indicating almost coincidence between the observed and expected numbers, there is a suggestion that the sample investigated was not randomly drawn from the underlying population.

Further values of T which, for a given ν, would be exceeded in f% of random samples (the f% confidence level), are tabulated as a function of f in Chapter 26 of Abramowitz and Stegun (1965).

A.4 Real Matrices

A number, N, of vectors $\boldsymbol{x}_i$ are said to be *linearly independent* if there are no linear relationships between them. N linearly independent vectors $\boldsymbol{x}_i$ are said to span N-dimensional space, in the sense that any other vector $\boldsymbol{a}$ in the space can be expressed as a linear combination of the $\boldsymbol{x}_i$, that is, $\boldsymbol{a} = \Sigma_i\, a_i\boldsymbol{x}_i$. Linear independence of the $\boldsymbol{x}_i$ ensures that the coefficients a_i are unique. For example,

$$\boldsymbol{x}_1 = \begin{pmatrix}2\\1\\0\end{pmatrix} \qquad \boldsymbol{x}_2 = \begin{pmatrix}0\\1\\2\end{pmatrix} \qquad \boldsymbol{x}_3 = \begin{pmatrix}0\\1\\0\end{pmatrix}$$

are linearly independent and we may decompose the vector

$$\begin{pmatrix}4\\6\\6\end{pmatrix} = 2\times\begin{pmatrix}2\\1\\0\end{pmatrix} + 3\times\begin{pmatrix}0\\1\\2\end{pmatrix} + \begin{pmatrix}0\\1\\0\end{pmatrix}.$$

A mutually orthogonal set of column vectors $\boldsymbol{a}_i$ satisfy the condition $\boldsymbol{a}_i^{\mathrm{T}}\boldsymbol{a}_j = 0$ where $\boldsymbol{a}_i^{\mathrm{T}}$ is the row vector obtained by transposing $\boldsymbol{a}_i$. A set of vectors is called *orthonormal* if it is both orthogonal and normalised:

$$\begin{aligned}\boldsymbol{a}_i^{\mathrm{T}}\boldsymbol{a}_j = \delta_{ij} &= 0 \qquad \text{if } i \neq j\\ &= 1 \qquad \text{if } i = j.\end{aligned} \tag{A.21}$$

(δ_{ij} is called the 'Kronecker' delta.) Orthogonal vectors are necessarily linearly independent. For example, the vectors

$$\begin{pmatrix}2\\0\\0\end{pmatrix} \quad \begin{pmatrix}0\\1\\1\end{pmatrix} \quad \begin{pmatrix}0\\1\\-1\end{pmatrix}$$

are orthogonal but not normalised.

Given a set of N linearly independent vectors, it is possible to construct from them a set of N orthogonal vectors spanning the same space using a step by step process called *Schmidt orthogonalisation*. If the $\boldsymbol{x}_i$ are linearly independent, then orthogonal $\boldsymbol{a}_i$ are constructed as follows:

$$\begin{aligned} \boldsymbol{a}_1 &= \boldsymbol{x}_1 \\ \boldsymbol{a}_2 &= \boldsymbol{x}_2 - (\boldsymbol{a}_1^{\mathrm{T}}\boldsymbol{x}_2)\boldsymbol{a}_1 \\ \boldsymbol{a}_3 &= \boldsymbol{x}_3 - (\boldsymbol{a}_1^{\mathrm{T}}\boldsymbol{x}_3)\boldsymbol{a}_1 - (\boldsymbol{a}_2^{\mathrm{T}}\boldsymbol{x}_3)\boldsymbol{a}_2. \end{aligned} \tag{A.22}$$

The *trace* (or spur) of a matrix is the sum of its diagonal elements. Hence, if

$$\mathbf{A} = \begin{bmatrix} 3 & 1 & 5 \\ 0 & 4 & 3 \\ 6 & 2 & 2 \end{bmatrix}$$

the trace, written Tr, is given by $\mathrm{Tr}\,\mathbf{A} = 3 + 4 + 2 = 9$.

It is often convenient, in algebraic manipulations, to partition a matrix into submatrices, for example,

$$\mathbf{A} = \begin{bmatrix} A_1 & A_2 \\ A_3 & A_4 \end{bmatrix} = \begin{bmatrix} A_1 & 0 \\ 0 & A_4 \end{bmatrix} + \begin{bmatrix} 0 & A_2 \\ A_3 & 0 \end{bmatrix}.$$

Symmetrical matrices are equal to their transpose: $\mathbf{A}^{\mathrm{T}} = \mathbf{A}$. *Antisymmetrical* matrices are equal to the negative of their transpose: $\mathbf{B}^{\mathrm{T}} = -\mathbf{B}$.

Any square matrix may be separated into a sum of symmetrical and antisymmetrical parts, for example,

$$\begin{bmatrix} 1 & 5 & 1 \\ -1 & 3 & -6 \\ 7 & -4 & 3 \end{bmatrix} = \begin{bmatrix} 1 & 2 & 4 \\ 2 & 3 & -5 \\ 4 & -5 & 3 \end{bmatrix} + \begin{bmatrix} 0 & 3 & -3 \\ -3 & 0 & -1 \\ 3 & 1 & 0 \end{bmatrix}$$

A square matrix $\mathbf{U}$ is said to be *orthogonal* if it satisfies the relation $\mathbf{U}^{\mathrm{T}}\mathbf{U} = \mathbf{I}$, where $\mathbf{I}$ is the unit matrix. In other words, the transpose of $\mathbf{U}$ is its inverse: $\mathbf{U}^{-1} = \mathbf{U}^{\mathrm{T}}$. The condition $\mathbf{U}^{\mathrm{T}}\mathbf{U} = \mathbf{I}$ can clearly be obtained if the n columns of $\mathbf{U}$ are constructed from n orthonormal vectors. Common examples of orthogonal matrices are the rotation matrices. For example, a rotation about the z axis of the vector

$$\begin{pmatrix} x \\ y \\ z \end{pmatrix}$$

through an angle θ corresponds to the matrix

$$\mathbf{M}(\theta) = \begin{bmatrix} \cos\theta & \sin\theta & 0 \\ -\sin\theta & \cos\theta & 0 \\ 0 & 0 & 1 \end{bmatrix}.$$

This clearly satisfies $\mathbf{M}^{\mathrm{T}}\mathbf{M} = \mathbf{I}$ for all θ.

Similarity transformations have the form $\mathbf{X} \rightarrow \mathbf{S}^{-1}\mathbf{X}\mathbf{S}$. They have several interesting properties. One such property is that the transform of a product is the product of transforms. A similarity transformation does not alter the trace of a matrix. The easy way to prove this (and many other results) is to use suffix notation:

$$\mathrm{Tr}(\mathbf{S}^{-1}\mathbf{A}\mathbf{S}) = \sum_{ijk} S_{ij}^{-1} A_{jk} S_{ki} = \sum_{ijk} (S_{ki} S_{ij}^{-1}) A_{jk}$$

$$= \sum_{jk} \delta_{kj} A_{jk} = \sum_{k} A_{kk} = \mathrm{Tr}\,\mathbf{A}. \qquad \text{(A.23)}$$

Orthogonal transformations are a special case of similarity transformations for which $\mathbf{S}$ is orthogonal. Hence, they can be written $\mathbf{X} \rightarrow \mathbf{U}^{\mathrm{T}}\mathbf{X}\mathbf{U}$ where $\mathbf{U}^{\mathrm{T}}\mathbf{U} = \mathbf{I}$.

The solutions of the eigenvalue equation $\mathbf{A}\boldsymbol{x}_i = \lambda_i \boldsymbol{x}_i$ define a set of eigenvectors, $\boldsymbol{x}_i$, and their corresponding eigenvalues, λ, of the matrix $\mathbf{A}$. An alternative way of writing the eigenvalue equation is $(\mathbf{A} - \lambda_i \mathbf{I})\boldsymbol{x}_i = 0$. This equation has the trivial solution $\boldsymbol{x}_i = 0$ if $(\mathbf{A} - \lambda_i \mathbf{I})$ has an inverse. Hence, the condition for non-trivial solutions is for the determinant $|\mathbf{A} - \lambda_i \mathbf{I}| = 0$, so that $(\mathbf{A} - \lambda_i \mathbf{I})$ has no inverse. The eigenvalues of $\mathbf{A}^{-1}$ are the reciprocals of the eigenvalues of $\mathbf{A}$ (the proof is trivial). Eigenvalues are invariant under similarity transformations, as can be seen by considering the following transform. Let

$$\mathbf{A} \rightarrow \mathbf{S}^{-1}\mathbf{A}\mathbf{S} = \mathbf{A}'.$$

Suppose

$$\mathbf{A}\boldsymbol{x} = \lambda \boldsymbol{x},$$

then

$$\mathbf{S}^{-1}\mathbf{A}\boldsymbol{x} = \lambda \mathbf{S}^{-1}\boldsymbol{x} \quad \text{or} \quad (\mathbf{S}^{-1}\mathbf{A}\mathbf{S})(\mathbf{S}^{-1}\boldsymbol{x}) = \lambda(\mathbf{S}^{-1}\boldsymbol{x}). \qquad \text{(A.24)}$$

Hence, $\mathbf{S}^{-1}\boldsymbol{x}$ is the corresponding eigenvector of $\mathbf{A}'$ which gives the same eigenvalue λ.

The determinant of a matrix product is equal to the product of determinants:

$$|\mathbf{A}\mathbf{B}| = |\mathbf{A}|\,|\mathbf{B}| = |\mathbf{B}|\,|\mathbf{A}|.$$

A simple proof of this result can be obtained by considering determinants of the partitioned product

$$\begin{bmatrix} \mathbf{I} & \mathbf{A} \\ \mathbf{0} & \mathbf{I} \end{bmatrix} \begin{bmatrix} \mathbf{A} & \mathbf{0} \\ -\mathbf{I} & \mathbf{B} \end{bmatrix} = \begin{bmatrix} \mathbf{0} & \mathbf{A}\mathbf{B} \\ -\mathbf{I} & \mathbf{B} \end{bmatrix}$$

but don't assume the result to be proved! The first matrix on the left-hand side merely produces linear combinations of the rows of the second matrix, and hence does not change the value of its determinant.

It follows that the determinant of a unitary matrix is unity. Also, determinants are invariant under similarity transformations of the matrix:

$$|\mathbf{A}'| = |\mathbf{S}^{-1}\mathbf{A}\mathbf{S}| = |\mathbf{S}^{-1}|\,|\mathbf{S}|\,|\mathbf{A}| = |\mathbf{S}^{-1}\mathbf{S}|\,|\mathbf{A}| = |\mathbf{A}|. \tag{A.25}$$

A real symmetric matrix has orthogonal eigenvectors for unequal eigenvalues; for $\lambda_i \neq \lambda_j$,

$$\mathbf{A}\boldsymbol{x}_i = \lambda_i \boldsymbol{x}_i$$

$$\mathbf{A}\boldsymbol{x}_j = \lambda_j \boldsymbol{x}_j \rightarrow \boldsymbol{x}_j^{\mathrm{T}}\mathbf{A} = \lambda_j \boldsymbol{x}_j^{\mathrm{T}}.$$

Hence, we can construct two expressions for $\boldsymbol{x}_j^{\mathrm{T}}\mathbf{A}\boldsymbol{x}_i$, giving

$$\lambda_i \boldsymbol{x}_j^{\mathrm{T}}\boldsymbol{x}_i = \lambda_j \boldsymbol{x}_j^{\mathrm{T}}\boldsymbol{x}_i.$$

Therefore, $\boldsymbol{x}_j^{\mathrm{T}}\boldsymbol{x}_i = 0$.

The normalised orthogonal eigenvectors of $\mathbf{A}$ may be collected to form an orthogonal matrix which 'diagonalises' it. Define $\mathbf{U} = \{\boldsymbol{x}_1\boldsymbol{x}_2\boldsymbol{x}_3 \ldots \boldsymbol{x}_N\}$, so that $\mathbf{U}^{\mathrm{T}}\mathbf{U} = \mathbf{I}$ and let $\mathbf{AU} = \{\lambda_1\boldsymbol{x}_1, \lambda_2\boldsymbol{x}_2, \lambda_3\boldsymbol{x}_3 \ldots \lambda_N\boldsymbol{x}_N\}$, then

$$\mathbf{U}^{\mathrm{T}}\mathbf{A}\mathbf{U} = \begin{bmatrix} \lambda_1 & & 0 \\ & \lambda_2 & \\ 0 & & \lambda_N \end{bmatrix} \tag{A.26}$$

Hence, the eigenvalue problem for a symmetric matrix is equivalent to the diagonalisation problem. Taking this with (A.25) above, we see that the determinant of $\mathbf{A}$ is equal to the product of all its eigenvalues. Taking (A.26) with (A.23), we see that the trace is equal to the sum of the eigenvalues. Similarly, (A.25) shows that the determinant of a matrix is equal to the product of its eigenvalues. Hence, result (A.24) really contains results (A.23) and (A.25).

More details about matrix methods can be found in Heading (1958).

References

Abramowitz M and Stegun I A ed 1965 *Handbook of Mathematical Functions* (New York: Dover)

Brandt S 1976 *Statistical and Computational Methods in Data Analysis* (Amsterdam: North-Holland)

Daniell G J, Hey A J G and Mandula J E 1984 *Phys. Rev.* D **30** 2230–2

Eadie W T, Drijard D, James F E, Roos M and Sadoulet B 1971 *Statistical Methods in Experimental Physics* (Amsterdam: North-Holland)

Heading J 1958 *Matrix Theory for Physicists* (London: Longmans)

Subject Index

Author Index

DISC ORDERING INSTRUCTIONS

Also available is a software package to accompany Computational Techniques in Physics. This is for use with selected projects and contains programs in BASIC and TURBO PASCAL, as well as a graphics program and data files. The software is suitable for running on IBM-PC microcomputers and compatibles and comprises a $5\frac{1}{4}$ inch double-sided, double density 360 kB floppy disc which operates on DOS 3.0 and higher versions.

Please use the order forms on the following pages to order the disc. Residents of the USA and Canada please use the correct form to ensure speedy delivery.

COUNTRIES OUTSIDE USA AND CANADA

ORDER FORM

SEND To: Adam Hilger
c/o IOP Publishing Ltd
Distribution Centre
7 Great Western Way
Bristol BS1 6HE
UK

☐ Please supply ________ copy/copies of the software disc (0-85274-429-3) for Computational Techniques in Physics at £15.00 each (£14.34 plus £0.66 VAT)

☐ I enclose my cheque for £________ made payable to IOP Publishing Ltd

☐ Please charge to my Access/Visa/American Express/Diners Club

Card No ☐☐☐☐☐☐☐☐☐☐☐☐☐☐☐☐ Expiry date /

Name ____

Address ____

____

____

Signature ____________________ Date ____________________

Telephone: (0272) 292151
Telex: 449907

USA AND CANADA ONLY

ORDER FORM

Mail to: **TAYLOR & FRANCIS**
International Publishers
242 Cherry Street
Philadelphia, Pennsylvania 19106-1906

To phone order call **215-238-0939**
Or Toll Free **800-821-8312**

MAILING INFORMATION

BILL TO:

Name ____________________

Address ____________________

____

____

City __________ State __________ Zip __________

ATTN ____________________

Phone (__________) ____________________

(Required for credit card payment)

SHIP TO:

Name ____________________

Address ____________________

____

____

City __________ State __________ Zip __________

ATTN ____________________

Phone (__________) ____________________

Orders from Individuals must be accompanied by payment or have credit card authorization.

Orders from Institutions must have PO No. and complete title information. Returns must be authorized in advance.

PAYMENT OPTIONS

Please prepay using one of these options

Make checks payable to TAYLOR & FRANCIS (US currency only).

1. Payment enclosed @ $__________ Check/Money Order # __________
2. Charge VISA __________ MC __________ AMEX __________

Card # ____________________ Exp. __________

Signature (required for credit card) ____________________

3. PO No. __________ DATE __________

SPECIAL INSTRUCTIONS ____________________

Shipping and Handling charges are $1.75 for the first book and $.50 for each additional. Or, if the order totals $50 or more, add 4%.

ORDERING INFORMATION

QTY	ISBN	TITLE	HC/SC $
	0-85274-429-3	Disc for Computational Techniques in Physics	$30.00 each

Prices subject to change without notice

MAIL TO:
TAYLOR & FRANCIS ● 242 CHERRY ST
PHILADELPHIA, PA 19106-1906
215-238-0939 ● **TOLL FREE 800-821-8312**

Subtotal @ __________

Shipping/Handling (see above) @ __________

6% Sales Tax (PA residents) @ __________

TOTAL @ **$** __________